False Feathers

Debora Weber-Wulff

False Feathers

A Perspective on Academic Plagiarism

 Springer

Debora Weber-Wulff
Berlin, Germany

ISBN 978-3-662-51393-4   ISBN 978-3-642-39961-9 (eBook)
DOI 10.1007/978-3-642-39961-9
Springer Heidelberg New York Dordrecht London

Printed on acid-free paper

Springer is part of Springer Science+Business Media (www.springer.com)

The Jay and the Peacock

A Jay venturing into a yard where Peacocks used to walk, found there a number of feathers which had fallen from the Peacocks when they were moulting. He tied them all to his tail and strutted down towards the Peacocks. When he came near them they soon discovered the cheat, and striding up to him pecked at him and plucked away his borrowed plumes. So the Jay could do no better than go back to the other Jays, who had watched his behaviour from a distance; but they were equally annoyed with him, and told him

"It is not only fine feathers that make fine birds."

Æsop, translated by Joseph Jacobs (1894)
The Fables of Æsop Selected, Told Anew and Their History Traced. London: Macmillan. p. 55.

Preface

As I was getting new glasses in May 2013, the optician saw my doctoral degree in his database and joked: "Do you still have a doctorate? One must ask these days."

When I began talking about plagiarism in Germany in 2002, it was a non-issue. It did not happen here, and so there was no need to talk about it at all. People did not understand why I, a professor of computer science, was speaking about plagiarism. I started testing plagiarism detection software in 2004 to see if it was effective. It wasn't. Sure, it could find a few sources, but it was not the magic litmus test for plagiarism that people were expecting. Beyond discussions about software, people were not interested in talking about plagiarism. There was (and still is) no widely accepted definition in Germany of what constitutes plagiarism. But now, in 2013, even the *Stammtisch* patrons have an opinion about plagiarism, especially by politicians and in doctoral dissertations. What happened?

In February 2011, the German Defense Minister Karl-Theodor zu Guttenberg was found to have heavily plagiarized in his doctoral dissertation, sparking a nation-wide discussion on the topic. When the first documentation platforms GuttenPlag Wiki and VroniPlag Wiki started up, I first stood on the outskirts, trying to explain to journalists both what this "swarm" phenomenon meant and what constitutes plagiarism. As more and more plagiarism was documented, I was drawn into the vortex and began participating in the documentation using my Wikipedia nick *WiseWoman*; I have been active ever since. The discussions with so many people from so many different fields about plagiarism have been quite fruitful, although unfortunately not much has actually happened yet in the way of changes to the German university system.

A book was needed on the topic, and although there are many fine books in English, they are not well known in Germany. I was planning on translating Jude Carroll's most excellent book on dealing with plagiarism at universities, *A Handbook for Deterring Plagiarism in Higher Education*, into German. I was going to include a chapter on zu Guttenberg and adapt her material to the German university system in order to get the word out that it is necessary that action be taken.

But her book focuses only on the student plagiarism problem and it has become increasingly clear that plagiarism is a systemic problem and is not just restricted to

Germany. There are people at all levels of academia in Germany and elsewhere who apparently either do not know how to properly use the materials of others or are willing to take shortcuts.

The name of this book, *False Feathers – A Perspective on Academic Plagiarism*, is based on the German-language e-learning unit "Falsche Federn Finden" (http://plagiat.htw-berlin.de/ff) that I produced during my sabbatical in 2004. The book is not, however, a translation of that material, which has become quite dated. I have included some historical material that has come to light in recent months during my sabbatical work in 2013, as well as an overview of how plagiarism is dealt with in some other countries. There is also much interest in how the crowd-sourced German project VroniPlag Wiki finds and documents plagiarism, so I have included a section on this topic.

Since there is now international interest in the question of plagiarism in Germany, it then seemed best to write this book in English, even if the target group is primarily German academics. Due to all of the public discussion about plagiarism, they are finding themselves in need of answers to three pressing questions: What exactly is plagiarism? How do we find it with as little effort as possible? And what do we do when we find it?

Not only do I find the problem of plagiarism interesting as a research subject, but I am also convinced that both plagiarism and the much wider and more serious problem of scientific misconduct in general are shaking the very basis of scientific investigation. I hope that this book can make a small contribution to combating this enormous problem by offering educators a good look into the various aspects of the plagiarism problem and suggesting some plans for finding answers to the questions above.

Like every book, this is the product of the work of many people, although only my name is actually on the cover. I would like to thank so many people by name for innumerable fruitful discussions, the careful reading of the many manuscript versions, and especially the hard work in digging out the historical plagiarism cases. I am indebted to you for your contributions and would have loved to include names here, but many have requested that I not even mention their pseudonyms. So let this be a global thank you for all who have closely accompanied the creation of this book. All errors are the result of me not following your excellent advice.

For friends and family who thought that I have been rather glued to my computer these past months: Thank you for your patience!

I am looking forward to reactions to this contribution to the discussion and welcome emails and letters.

Berlin, Germany   Debora Weber-Wulff
September 2013   *weberwu@htw-berlin.de*

Contents

List of Figures

List of Tables

Chapter 1
Introduction

Since human beings have been writing, there has been plagiarism, as can be seen in some excellent histories of plagiarism, for example Mallon (2001) or Theisohn (2009). It is not something that sprouted with the advent of the Internet. Teachers have been struggling for years in countries all over the globe to find good methods for dealing with the problem of students who plagiarize. When do we suspect plagiarism? How do we judge when a text is to be considered plagiarism, since it is not a binary yes/no decision? How do we teach students not to plagiarize? And how do we deal with those who have been caught plagiarizing?

The purpose of this book is to collect material on the various aspects of plagiarism in education, with special attention given to the problem of plagiarism in dissertations in Germany.

An overview of plagiarism definitions will be given as an attempt to clearly differentiate between plagiarism and the other forms of scientific misconduct that are often lumped under the heading of plagiarism. Since there are many different kinds of plagiarism beyond simple copy & paste, a typology of plagiarism will also be set out.

In Germany, not only has there been some discussion surrounding plagiarism by students, but since early 2011 many dissertations have been documented as containing extensive plagiarism. It was then that plagiarism in the dissertation of the Minister of Defense, Karl-Theodor zu Guttenberg, was documented by a so-called "swarm" of unknown persons. This phenomenon of public plagiarism documentation as carried out at the GuttenPlag Wiki and the VroniPlag Wiki will be presented in detail, as well as some information about historical plagiarism in dissertations and other academic works at German-speaking universities, thereby demonstrating that plagiarism – even in dissertations – is really nothing new.

On the question of plagiarism detection, this book will try to explain in non-technical terms how plagiarism detection systems work, and why they are just partial copy-detection systems. Reports from up-to-date tests of the software are included in Sect. 4.1.2. Software systems can be useful tools, but they cannot in and of themselves determine plagiarism – that is something that educators must determine themselves. It is, however, more efficient and more effective to use a search engine in order to discover probable sources. A detailed how-to guide for finding plagiarism using just simple tools will be given. Appendix F collects the URLs of the many systems and sites that are discussed in this book; the site names are given in italics in the text.

This book describes in detail how the VroniPlag Wiki group works in discovering and documenting plagiarism in dissertations. Journalists often assume that there is some sort of industrial-strength software involved, when it is just the tool-based collaborative and interactive application of traditional research methods and modern digital technology and databases to the question of uncovering and documenting plagiarism. Collusion, a small subset of plagiarizing practice, is, however, somewhat detectable using software systems and will thus be discussed in Sect. 4.5.

The focus of the efforts on the part of the universities must, however, be in plagiarism avoidance. Some possible methods for combating plagiarism will be given, as well as policies and procedures on plagiarism from a selection of countries.

This book will not be able to address the myriad legal issues that plagiarism entails, even though they may be briefly mentioned. It also cannot present a one-size-fits-all answer to the question of how to deal with plagiarism at a particular university. But it should give a thorough overview of the various aspects of the questions. The perspective outlined in Chap. 5 involves training students and teachers, having a transparent policy, and testing a sample of theses. It is hoped that such an approach would begin to address this grave problem. From this, instructors and universities should be able to develop their own, custom-fit policies for dealing with a serious problem: using others' words or ideas without attribution.

References

Mallon, T. (2001). *Stolen Words*. (2nd ed.). San Diego: Harcourt.
Theisohn, P. (2009). *Plagiat: eine unoriginelle Literaturgeschichte*. Stuttgart: Kröner.

Chapter 2
Plagiarism and Academic Misconduct

If one wants plagiarism and academic misconduct to be addressed fairly and consistently, there must be good definitions available that are more or less universally agreed upon. This is where the trouble begins, as there are numerous definitions in English that focus on different aspects of the problem. In Germany, however, there is currently no legal definition of plagiarism, in particular because the responsibility for universities lies with the power of the states, not with the federal government. The states, in turn, tend to delegate responsibility for such topics to the universities. Each university will, if attempting to define it at all, have its own definition of the term. So in the absence of a legal definition, any plagiarism cases that end up in court force the judges to look at prior law on the topic instead of being able to refer to an codified definition.

This chapter will be concerned with presenting and discussing some of the various definitions for plagiarism and academic misconduct, as well as examining the various types of plagiarism. A section on the magnitude of the problem and the reasons given by students for cheating is followed by a discussion of the reasons that plagiarism is a problem.

2.1 Definition of Plagiarism

The word *plagiarism* is derived from the Latin word *plagiarius*, meaning someone who kidnaps the child or slave of another. A Latin poet, Marcus Valerius Martialis, used this term in *Epigrams 1, 52* to describe a fellow poet, Quintianus, who published Martial's poems as his own. Martial felt that his poems were the children of his mind, and that they were being stolen. The term, however, only began being commonly used to describe what is sometimes also called literary theft around 1600.

There is a rather joking definition of plagiarism that Stuart McIver attributes to Wilson Mizner (McIver 1994, p. 66):

> If you copy from one author, it's plagiarism;
> if you copy from two, it's research.

A definition that is often used in the United States was first developed by Joseph Gibaldi for the *MLA Style Manual and Guide to Scholarly Publishing*, published by the Modern Language Association (Gibaldi 1998, 2003), and is now given in the third edition (Modern Language Association 2008, p. 166) as:

> [T]he act of plagiarism gives the impression that you wrote or thought something that you in fact borrowed from someone, and to do so is a violation of professional ethics.
>
> Forms of plagiarism include the failure to give appropriate acknowledgment when repeating another's wording or particularly apt phrase, paraphrasing another's argument, and presenting another's line of thinking.

Teddi Fishman, the director of the International Center for Academic Integrity, proposed a definition of plagiarism that includes many important aspects in (Fishman 2009, p. 5):

> Plagiarism occurs when someone
>
> 1. Uses words, ideas, or work products
> 2. Attributable to another identifiable person or source
> 3. Without attributing the work to the source from which it was obtained
> 4. In a situation in which there is a legitimate expectation of original authorship
> 5. In order to obtain some benefit, credit, or gain which need not be monetary[.]

In Germany there is neither a generally agreed upon definition for plagiarism, nor is there a codified legal definition to be found. One definition was given by the author Paul Englisch (1933, p. 81–82):

> Plagiat ist also die aus freier Entschließung eines Autors oder Künstlers betätigte Entnahme eines nicht unbeträchtlichen Gedankeninhalts eines anderen für sein Werk, in der Absicht, solche Zwangsanleihe nach ihrer Herkunft durch entsprechende Umgestaltung zu verwischen und den Anschein eigenen Schaffens damit beim Leser oder Beschauer zu erwecken.
>
> Plagiarism is the conscious decision of an author or artist to take a not insignificant portion of the contents of the thoughts of someone else for his own work with the intention of disguising the origin of the forced loan by rearrangement, so as to arouse in the reader or observer a pretense of original production.
> *(translation by the author)*

As it turns out, Englisch was himself accused of plagiarism and apparently was trying to set up a definition that made his own use of another's text appear more acceptable. More on the Englisch plagiarism case will be given in Sect. 3.5.4.

Mizner's definition, although poking fun at research, does actually hold a kernel of truth as some people truly believe that research consists of compiling what other people have said. By taking text from multiple places and making a change here or there, putting the reference in the bibliography, they feel that they have produced something of their own. It is also ironically interesting to see that this definition may not actually be from Mizner. Garson O'Toole (2010) discusses alternate possible origins for the same saying at length.

Gibaldi's definition makes it clear that not just a word-for-word copy is to be considered plagiarism, but that also a good phrase, and of course a paraphrase or even the re-use of structural arguments, can be considered to be plagiarism if the source is not given. Gibaldi also makes it clear that this is an ethical question, not one of copyright as is often assumed.

Fishman's definition is quite comprehensive. It specifies that not only copyright infringement or a complete copy is to be considered as constituting plagiarism. The definition also clarifies that expressing general ideas and facts are not plagiarism when they are not attributable to a specific person or group of persons. And we only speak of plagiarism in situations where there is an expectation of original work. You can copy word-for-word any texts you like for private use, but if you submit them as your own work in a situation in which you are expected to be doing your own work, then it would be considered plagiarism, although not everyone agrees with this. There is a bit of debate about whether a text should be considered a plagiarism in and of itself, or whether the circumstances (private vs. public use, intent) should have a role to play in the determination of whether or not it is to be considered plagiarism. In particular, the question needs to be weighed as to whether the act of publishing a text that has been a private copy suddenly makes the text a plagiarism, or if making a private copy without attribution is in and of itself a plagiarism that may, however, not be seen as problematic in a moral sense.

The author of this book had been using Englisch's definition for many years in Germany, as it is rather elegantly written. It is problematic that he includes intent here, but without knowing the story behind the definition it did not seem to be that important. However, as it turns out, Englisch's insistence that only text taken on purpose is plagiarism is certainly designed to absolve himself of being accused of plagiarism. Interestingly, much of the discussion in Germany since 2011 has also centered around this question of intent. Is it only plagiarism if it is done deliberately? This does raise the question of proof – since intent is a state of mind, it is difficult to prove that something was done with intent and did not just happen accidentally. However, if minor changes are made to a text, it can be seen that it was not just a case of forgetting to set the quotation marks, but that the text must have been copied and changed by a conscious decision.

Another possibility for demonstrating intent is when references are copied from the source and the source itself contained bibliographic errors. If these errors also appear in the copied text, it is clear that this author did not literally go and look at the source in question, but just copied the reference. This can additionally be seen as intent to deceive, as the author is pretending to have done research that was not actually done. But even if the reason for the plagiarism is just bad working habits and not a brazen attempt to deceive – it is still, in the author's opinion, plagiarism. The consequences might be different depending on how clearly intent is perceived to have been involved, but nevertheless the text has indeed been copied.

Gibaldi, Fishman, and Englisch all address the notion of idea or structural plagiarism. This is a difficult issue, as will be discussed in Sect. 2.2.6, because it is entirely possible that people can come up with the same ideas independently of each other. But when the sequence of arguments or selection of illustrations extends over pages and chapters, the plagiarism becomes much clearer.

The question of accidentally writing an identical text is also often raised in connection with plagiarism definitions. It seems logical to many people that there are only so many ways to state certain things, and thus they find it highly possible that text passages can be replicated without intent. There are, however, countless forms for expressing a thought, and even the description of a scientific method can be varied without distorting the meaning. And should it be necessary to reuse a text from a previous publication, this is of course possible if the author gives the reference, i.e. makes clear where the text first appeared.

The Fishman definition is the one that, in the opinion of the author, should be adopted for university use. Even though the definition is extensive, it does rather nicely capture many aspects of what constitutes plagiarism. In particular, this definition makes it clear that the plagiarist is benefiting personally from taking the text. It might just mean obtaining a few more points or being awarded credits, but it might even be the final thesis that leads to graduating with a degree. No matter at what level the plagiarism takes place, this gives the plagiarist an unfair advantage over honest students and researchers. And the plagiarist may continue to profit from this deed, for example if a dissertation is plagiarized and a well-paying job is obtained on the basis of the doctorate.

But what about a situation in which someone copies a chapter from a book into a Wikipedia article? This could be seen as not violating the fifth point in Fishman's definition. But if we observe that this situation would be a two-fold gain, an increase in the edit count for the Wikipedia editor and an increase in the amount of information found in the encyclopedia itself (both of which are not monetary gains), then it is clear that the definition also covers exactly this case.

Of course, there is a difference between determining plagiarism and deciding the consequences an act of plagiarism is to have. This is an issue that every school and university will have to determine for itself. Not every kind of plagiarism will have the same, perhaps quite severe, sanctions leveled. Chapter 6 will give a few examples from other countries. But it is important that both the definition and the sanctions at a particular institution are clearly communicated to all students and researchers.

Some people also mistakenly equate plagiarism with copyright violations. While it is certainly true that some instances of plagiarism may violate the copyrights of the original authors, and some copyright violations are themselves plagiarisms, plagiarism also includes taking text that is not or no longer under copyright, or even material that one has written oneself, and using it without appropriate quotation. This is not good academic practice, even if it could be considered to be a legal use of the material. On the other hand, five pages that are correctly quoted could also be considered a copyright violation, as the quotation far exceeds what would normally be considered fair use. For an extensive discussion of the copyright implications of plagiarism, see Dreier & Ohly (2013).

2.2 A Typology of Plagiarism

The definitions above do not make it clear that there are also various forms of plagiarism above and beyond simplistic copy & paste. Attempts to classify types of plagiarism have been given by Lancaster (2003), Maurer, Kappe, & Zaka (2006), and Stein & Meyer zu Eissen (2007), among others, but these classifications are often about different aspects such as single vs. multiple source or intrinsic vs. extrinsic plagiarism.

One important form of plagiarism that can often be found in countries with a national language other than English is translations. Paraphrasing can also be broken down into different forms, and there are specific kinds of plagiarism that have only rather recently been named. This chapter discusses an extended typology of plagiarism that was first proposed by Weber-Wulff & Wohnsdorf (2006).

2.2.1 Copy & Paste

This is the easiest kind of plagiarism to create – a portion of online text (or the entire paper) is marked and with two double keystrokes (CTRL + C and CTRL + V), a copy is made and inserted into another document. Since the passages are identical, this kind of plagiarism is relatively easy to find. The sequence of characters, including spaces, punctuation marks, and spelling errors, is the same in the original and in the copy, although often a few minor edits are made in order to disguise the copy.

Indeed, it is often surprisingly simple to detect this kind of plagiarism. An alert teacher will spot strange errors or a writing style that is far beyond the capabilities of the student. And since students who use copy & paste techniques often do not proofread the texts that they produce, there will be glaring errors, as the author has often experienced herself. They ignore inconsistencies in spelling words that switch between American English and British English or the old and new spelling in German, or the use of the letter 'ß' instead of 'ss' in Swiss German, and of course they generally preserve all misspellings. Either using plagiarism detection software or feeding a search engine just a few choice words will often discover some sources for this kind of plagiarism.

Sometimes unintentional portions of the original text are copied as well. When copying from the Internet, editors such as Word will copy the links along with the text, so that some words appear underlined in the plagiarism. Advertisements are also easily and inadvertently copied this way. When copying from a PDF, the page numbers and the page headings are also copied, so sudden numbers in the middle of a sentence or a repeated phrase that could be a section title or other artifacts found in strange places in a paper can be good indicators that a more detailed plagiarism investigation is warranted.

2.2.2 Translations

For a translation, the plagiarist chooses a text portion in a language different from the target language and either translates by hand or uses an online translation tool such as *Babelfish* or *Google Translate* to produce a rough draft. Many students are apparently not aware that such software does not produce a high-quality translation. A native speaker will quickly see that the text is in some way peculiar. Some will react to word order, others to strangely wrong word selections or to incorrect grammar. Putting some words that are identical in both languages, for example proper names, into a search engine can be instrumental in discovering the source for such translations.

A translation done by hand will be much more difficult to spot, especially if it is a good one. The author once had a student who used an automatic translator to produce a rough draft for a paper due for his English class. He handed it in and then his English instructor marked up the problems, which he fixed before handing the next version of the same paper on to me. Luckily, there was a misspelled word in the original and both the automatic translator and the English teacher thought this to be a technical term and let it stand. This word, together with some proper names, was helpful in finding the source.

Some students feel that through the work done in preparing a hand translation, effort has been expended and should be rewarded. Indeed, such a translation is a great deal of work, but it is not original work, as it is still entirely based on the work of someone else.

2.2.3 Disguised Plagiarism

Sometimes students ask questions like this (lol 2008): "How many words do you have to change in a document for it not to be plagiarised?" This notion that the ideas and texts somehow end up belonging to someone else just because a few words were changed or the word order switched or a phrase inserted seems to be quite widespread among students. Simply changing words around or inserting or deleting a phrase, however, does not result in an original work, but an edited work, and thus it is still a plagiarism, although it can be much harder to detect as it is no longer identical to the source character-wise.

Some systems are available online that will automatically and without cost exchange every sixth or seventh word with a synonym from a thesaurus. This will result in a quite unintelligible text that will easily confuse a plagiarism detection system into thinking that the text is not plagiarized. But it still remains a plagiarism, even though it is a disguised one.

2.2.4 Shake & Paste Collections

Another technique popular with students in the preparation of a plagiarized paper is to take a number of sources and to copy the material paragraph-wise (or even sentence by sentence) from various sources. The copied portions are just assembled one after another, often in no particular logical order. This tends to give the reader the impression that the paragraphs were put into a bag, shaken well, and taken out for pasting in a rather random fashion.

A human reader will often pick up on this type of plagiarism because of the jarring changes in style, diction, or even formatting from paragraph to paragraph, even though each paragraph by itself seems nicely written. Some research has been conducted in the area of intrinsic plagiarism analysis in the hopes of identifying such changes in a text, for example Stein & Meyer zu Eissen (2007).

Stein and Meyer zu Eissen calculate so-called style markers over the entire document, and then compute the style markers for individual paragraphs. If they differ by much, this could be an indication of possible plagiarism. Style markers are statistics such as the average length (in words) of sentences, the average number of adverbs or verbs or numbers or nouns or dashes in a text. The research group has also had some success using the average word frequency class or the pattern of the 250 most frequent words used. This technique, however, can only answer the question: Does this paper seem to have been written by one person or by several?

Such efforts are still at a very preliminary stage. It must be understood, however, that this method, called intrinsic plagiarism analysis, does not actually find any potential sources.

2.2.5 Clause Quilts

A clause quilt is a variation of paraphrasing plagiarism that has been called patchwriting by Rebecca Moore Howard. She defines patchwriting to be the "blending [of student's] words and phrases with those of the source – with or without acknowledgment" (Howard 1999, p. 117). Students take bits and pieces of text from different authors, stitch the half-sentences together and edit them, perhaps changing an adjective here or switching the word order there. Some glue nicely worded sentences together with their own, grammatically incorrect phrase or two in an attempt to emulate academic writing. The result is a crazy quilt of text fragments that can sometimes border on the meaningless. Some authors also use the term *mosaic plagiarism* for this type of plagiarism, e.g. Harvard University (2013).

Howard continues, noting (1999, p. 117) that it is often students who are writing in a foreign language that produce this kind of plagiarism, although she also suggests that teaching students to patchwrite on the basis of good academic texts will teach them how to write academically, an approach the author finds quite problematic.

It is easy to spot this kind of plagiarism, but it is harder to find the sources, as there may be quite a number. Documenting the situation is a daunting task, as each snippet taken for itself might just be considered to be borderline plagiarism, but taken together on page after page, it becomes clear that these are not the words of the student.

2.2.6 Structural Plagiarism

In a structural plagiarism, the plagiarist will paraphrase another author without giving credit. This can include using the argumentative structure, the sources (sometimes even in the exact order as in the original work), the footnotes, the experimental setup, or even the research goal. It is impossible to detect this kind of plagiarism with the help of software or even to automatically prove that it exists, given the two sources in question. However, structural plagiarism is often the basis of heated arguments between academics who insist that others have stolen their ideas.

Mediating such disputes is a daunting task, especially as it is entirely possible that two persons can come up with the same idea independently. But the chances of them using the exact same argumentative structure dwindles for every additional similarity. A structural plagiarism can, nevertheless, be handily proven if the later work makes the exact same errors as the work it is based on, or new errors that are only possible if the later work was available, for example erroneous literature references. One example from the area of mathematics that led to a mild sanctioning of the plagiarist is demonstrated by Gumm (2010).

2.2.7 Pawn Sacrifice

Benjamin Lahusen (2006) analyzed a book published by a Berlin law professor and in doing so identified a common type of plagiarism that he called a *pawn sacrifice*. This means that the source citation is either given in a footnote or only listed in the bibliography. It is not made clear, however, exactly how much has been taken – and often the passage is taken word for word. A variation of this is a proper attribution of a sentence, but then the text copy continues on, copying the source for additional sentences or even paragraphs without making clear that this is the author of the source speaking and not the purported author.

An interesting example of the pawn sacrifice can be seen in Fig. 2.1. Two paragraphs from the 2009 dissertation of Karl-Theodor zu Guttenberg (2009, p. 345) that are taken from a talk given by Gret Haller (2003) are shown in juxtaposition with the corresponding portion from Haller's online manuscript. The example was colored by hand.

Although this text is an almost exact copy, there are no quotation marks given, just the reference in a footnote to the first paragraph. In the second paragraph, zu Guttenberg continues to take text from Haller. One can see a few more changes: a comma followed by *und* twice became a semicolon; a filler word, *übrigens*, was inserted; and a *dafür* became a *hierfür*. Since the original author is from Switzerland and the Swiss do not use the *ß* letter, one word that uses the *ß* in other parts of the German-speaking world was correctly changed from *ss*. Additionally, Haller's use of a more gender-neutral language to speak about the writers of the constitution, *Gründern* (founders), is made explicitly male with *Gründerväter* and then the En-

Guttenberg, K.-T. zu (2009)
*Verfassung und Verfassungsvertrag: Konstitutionelle
Entwicklungsstufen in den USA und der EU.*
Berlin: Duncker & Humblot, p. 345

In Europa besteht „demokratische Identität" in der Wahl der
Parlamente, zu der man in der Eigenschaft als Teil des
Volkssouveräns berechtigt ist. US-Amerikaner erleben
demokratische Identität weniger in diesem Bereich als darin,
Rechte zu haben, auf die man sich jederzeit gerne zu berufen
vermag, und die man als Einzelperson oder Vertretung eines
Minderheitsinteresses vor Gericht einklagen kann. Demzufol-
ge erhalten Recht und Justiz in den Vereinigten Staaten eine
gänzlich andere Funktion als in Europa, nämlich letztlich eine
in weiten Teilen politische. [998]

In Europa bedeutet übrigens „Politik" unter anderem, dass in
den politischen Instanzen, insbesondere in den Parlamenten
um die Gesetzgebung gestritten wird; die so entstandene
Rechtsordnung wird dem Staat anvertraut. In den Vereinigten
Staaten wird um Rechte gestritten; der Staat schafft hierfür nur
den äußeren Rahmen. Wenn in den Vereinigten Staaten die
Auseinandersetzung um die Verteilung von Macht direkt –
horizontal – in der Gesellschaft zwischen den Privaten
stattfindet, und nur zu einem kleineren Teil im Parlament, so
deshalb, weil den Gründervätern dieser Nation die Vorstellung
eines „vernünftigen Gemeinwillens" fremd war, der in Europa
der Staatsbildung weitgehend zugrunde liegt. Die „founding
fathers" wollten eine möglichst staatsfreie Gesellschaft, in
welcher die Machtverteilung zwischen Privaten oder allenfalls
Minderheitsgruppen ausgehandelt wird, um Mehrheiten zu
vermeiden, welche die Legitimation hätten beanspruchen
können, den Staat zu stärken.

[998] Vgl. auch *G. Haller*, Recht – Demokratie – Politik. Zum
unterschiedlichen Verständnis von Staat und Nation dies- und
jenseits des Atlantiks. Referat anlässlich der Tagung „Die
USA – Innenansichten einer Weltmacht", 7./ 8. Februar 2003
an der Katholischen Akademie in Bayern, München,
http://www.grethaller.ch/kath-ak-muenchen.html.

Haller, G. (2003)
*Recht – Demokratie – Politik. Zum unterschiedlichen Ver-
ständnis von Staat und Nation dies- und jenseits des Atlantiks.*
`http://www.grethaller.ch/texte/2003/recht-
demokratie-politik_europa-und-usa.html`

In Europa besteht demokratische Identität in der Wahl der
Parlamente, zu der man in der Eigenschaft als Teil des
Volkssouveräns berechtigt ist. US-Amerikaner erleben
demokratische Identität viel weniger in diesem Bereich,
sondern darin, Rechte zu haben, auf die man sich jederzeit
gerne beruft, und die man als Einzelperson oder Vertretung
eines Minderheitsinteresses vor Gericht einklagen kann. So
erhält das Recht und die Justiz in den Vereinigten Staaten eine
ganz andere Funktion als in Europa, nämlich eine politische.
[...]

In Europa bedeutet Politik unter anderem, dass in den
politischen Instanzen, insbesondere in den Parlamenten um die
Gesetzgebung gestritten wird, und die so entstandene Rechtsordnung wird dem Staat anvertraut. In den Vereinigten Staaten
wird um Rechte gestritten, und der Staat schafft dafür nur den
äusseren Rahmen. Wenn in den Vereinigten Staaten die
Auseinandersetzung um die Verteilung von Macht direkt –
horizontal – in der Gesellschaft zwischen den Privaten
stattfindet, und nur zu einem kleineren Teil im Parlament, so
deshalb, weil den Gründern dieser Nation die Vorstellung
eines vernünftigen Gemeinwillens fremd war, der in Europa
der Staatsbildung zugrundeliegt. Sie wollten eine möglichst
staatsfreie Gesellschaft, in welcher die Machtverteilung
zwischen Privaten oder allenfalls Minderheitsgruppen
ausgehandelt wird, um Mehrheiten zu vermeiden, welche die
Legitimation hätten beanspruchen können, den Staat zu
stärken.

Fig. 2.1 Pawn sacrifice: zu Guttenberg (2009, p. 345) vs. Haller (2003)

glish equivalent, *founding fathers* (set off in quotation marks) is inserted into the
next sentence, replacing the pronoun.

But these minor changes did not make the text his own. There is no reference to
Haller besides the footnote on the first paragraph. In particular, there is no mention
that these are, in essence, her words, slightly modified.

Ingo von Münch (2012, pp. 123–126) does not accept that this constitutes plagiarism because it is apparently a very common type of reference in legal texts. He
only sees a plagiarism when the original author is not named, and with this kind of
reference the author is given. However, as can easily be seen in the zu Guttenberg
example, what is missing is a clear delineation of where the paraphrase starts and
where it ends, if it is indeed a paraphrase. A word-for-word copy needs quotation
marks and must not be changed, or the changes must be clearly indicated.

2.2.8 Cut & Slide

Another interesting type of plagiarism was first identified in the zu Guttenberg thesis, but has since been found in a number of other plagiarized dissertations. The author has dubbed this "cut & slide" plagiarism. In this type of plagiarism, a portion of the original text is degraded to only being a footnote or even moved to an appendix. Figure 2.2 shows an example from the zu Guttenberg dissertation. On the right in the figure is a passage from a newspaper article (Kühnhardt 2004), on the left at the top is a paragraph from zu Guttenberg's thesis, (2009, p. 189), and on the bottom the corresponding footnote from the following page. In this particular example, it is interesting to see that the portion slid into the footnote is properly quoted, the portion that remains in the main text, although equally word for word, is not given in quotation marks.

The source is given in this particular instance, so it is similar to a pawn sacrifice plagiarism. But it is given for only one part of the source, and the statements that the original author found to be so important that they were written together are now split apart with one part marginalized.

zu Guttenberg, K.-T. (2009)
Verfassung und Verfassungsvertrag: konstitutionelle Entwicklungsstufen in den USA und der EU. Berlin: Duncker & Humblot, pp. 189-190.

Seit dem Abschluss der Einheitlichen Europäischen Akte 1986 ist dies ein Indikator dafür geworden, dass die europäische Integration auf die Identität ihrer Mitgliedsstaaten zurückwirkt. Die Frage nach der konstitutionellen Legitimität einer vertieften Integration stellt sich überhaupt nur dort, wo der nächste politische Schritt tatsächlich eine Vertiefung des Integrationsprozesses bedeutet.[537]

[537] Vgl. *L. Kühnhardt*, Auf dem Weg zu einem europäischen Verfassungspatriotismus, in: NZZ, 16. Juli 2004: „Wo dies der Fall ist, geht es um die Übertragung nationalstaatlicher Souveränität auf die EU. Es ist nicht verwunderlich, dass in einer solchen Situation in einigen Ländern der EU die Referendumsfrage virulent wurde – und bei der europäischen Verfassung wieder virulent geworden ist. Andere Staaten votierten schon in früheren Fällen – und auch jetzt wieder – für die primäre Verantwortung ihrer frei gewählten und dadurch entsprechend zur Abstimmung mandatierten Parlamente."

Kühnhardt, L. (2004)
Auf dem Weg zu einem europäischen Verfassungspatriotismus. In: *Neue Zürcher Zeitung*, Vol. 225, No. 163, 16 July, p. 9.

Seit dem Abschluss der Einheitlichen Europäischen Akte 1986 ist dies ein Indikator dafür geworden, dass die europäische Integration auf die Identität ihrer Mitgliedsstaaten zurückwirkt. Die Frage nach der konstitutionellen Legitimität einer vertieften Integration stellt sich überhaupt nur dort, wo der nächste politische Schritt tatsächlich eine Vertiefung des Integrationsprozesses bedeutet. Wo dies der Fall ist, geht es um die Übertragung nationalstaatlicher Souveränität auf die EU. Es ist nicht verwunderlich, dass in einer solchen Situation in einigen Ländern der EU die Referendumsfrage virulent wurde – und bei der europäischen Verfassung wieder virulent geworden ist. Andere Staaten votierten schon in früheren Fällen – und auch jetzt wieder – für die primäre Verantwortung ihrer frei gewählten und dadurch entsprechend zur Abstimmung mandatierten Parlamente.

Fig. 2.2 An example of cut & slide plagiarism

2.2.9 Self-plagiarism

Many researchers are of the opinion that if they are the authors of a text, they own it, and thus they can reuse it as often as they like without citing the original publication. This is a typical confusion of copyright issues and matters of good academic practice.

Unless the author has signed away the reproduction rights to a text as part of a journal article or a book, the author of course has the legal right to reuse a text. But it is not good academic practice to force readers who are trying to gain an overview of a subject matter to read the same text over and over again. The simplest way to reuse a text is to mark it appropriately – beginning and end – and to reference the original source so that readers who have already read the source need not focus on trying to discover if there is any minor change to this portion of the writing, but can continue reading with what is ideally new material.

Submitting an identical or nearly identical paper to different conferences or journals, or reusing a paper with different author lists is clearly not acceptable in terms of good academic practice. If the reviewers do not catch on that this paper has already been published, or if it is submitted in parallel to multiple venues, then the paper might slip through the cracks and serve only to inflate the author's CV. Such a duplicate publication can only be tolerated as acceptable academic practice if both of the publications are aware of the duplication, and if the one to appear later includes a clear indication that this is just a copy republished in order to reach a wider or a different audience. Authors who play fast and loose with this form of publication may soon find themselves facing retractions, as publishers are starting to crack down on duplicate publication.

More information about self-plagiarism can be found in the online resource material Miguel Roig began developing in 2002 for the Office of Research Integrity in the USA. He addresses the self-plagiarism problem at length in guidelines published in Roig (2003). In particular, he discusses redundant or duplicate publication, data fragmentation issues that are sometimes referred to as "salami slicing", copyright infringement, and text recycling, especially as is often found in reusing method or experimental setup descriptions. Patrick M. Scanlon (2007) disagrees with Roig on the last point, finding that a limited amount of self-plagiarism should be permissible, a conclusion he draws after an extensive discussion and analysis of the literature on the topic.

2.2.10 Other Dimensions

There are other dimensions that could be used in classifying types of plagiarism. For example, some plagiarisms only take material from one source, while others use multiple sources. There can also be combinations of types, e.g. a shake & paste plagiarism could also entail a translation, or a clause quilt of self-plagiarisms could be constructed. And even though citations are given, the citation may be misplaced or intended to cover a quotation that extends over multiple pages, or even be completely missing with the source only given in the bibliography.

This typology does not define a discrete space for categorizing an entire work as a specific type of plagiarism. Rather, it can be used for attempting to classify individual fragments, keeping in mind that plagiarism is a continuous spectrum of text manipulations and not just one particular method of using other people's words.

Completely missing from this typology is the question of intent, as was mentioned in Sect. 2.1 while looking at definitions of plagiarism. It is usually impossible to tell from the text itself if the text parallel is the result of unintended, honest mistakes, sloppy referencing skills, or intent to cut corners and to deceive. There may not be a specific moment in which an author decides: Okay, now I'm plagiarizing. It may start with a text paraphrased here and a reference forgotten there, and then in the time crunch as the deadline looms, more and more material is appropriated. At some point – dependent on the type of text being written, the size of the text, and the amount and kind of plagiarism found – it just becomes unacceptable. This makes it impossible for there to be simple rules for categorizing or classifying plagiarisms, as that tipping point cannot be precisely described. There must always be a human reader who makes the decision that enough is enough.

2.3 Other Types of Academic Misconduct

People often confuse the notion of plagiarism with the more broader topic of academic misconduct, or they expand plagiarism to include minor errors such as forgetting to include the publisher in a reference. Since some serious behaviors are often wrongly subsumed under the heading of plagiarism, a brief discussion of them is given in this section. They are not generally amenable to plagiarism detection methods, and thus need to be specifically discussed with students in order to make them aware that this kind of conduct is not tolerable.

2.3.1 Ghostwriting

A ghostwriter is someone who writes texts for other people but does not claim authorship. Some politicians use ghostwriters, but in academia this practice is not acceptable. Students are supposed to be learning how to research and to write papers themselves and academics are assumed to be the true authors of documents bearing their names. There are many different kinds of ghostwritten papers available:

- *Custom-written papers* can be purchased for a fee from companies that hire writers to produce them to order. Even doctoral dissertations are available for a hefty price. There is often no direct contact between the ghostwriter and the person handing in the paper, as they transact via an intermediary company in order to preserve the anonymity of the ghostwriters.

- There are numerous *paper mills* that collect papers either with the explicit permission of the authors or that harvest them by more or less dubious methods. The papers are kept in a large, searchable database. Access is sold to the customer under the guise of showing them how a research paper is to be structured. Usually, a warning is given not to use the paper verbatim, but this warning is not always heeded.
- Since theses are available in many *university libraries* and they often have a CD with a digital version of the thesis included on the back cover, students from nearby universities have been known to spend time looking for interesting topics that they can then copy or adapt and submit as their own. They propose this topic for their thesis, and hand in a slightly modified version on the last day, turning the original authors into unwitting ghostwriters.

Judging from the number of companies advertising ghostwriting services and paper mills, it appears to be a widespread problem that is unfortunately difficult to combat. Many companies that advertise such services insinuate that there is really nothing wrong with getting a little "editing help" or "inspiration" for a paper. Of course, if the purchased paper was indeed just used as an inspiration, then there would be little to say about it. But when students or researchers submit papers that have been ghostwritten in their own name, then *they* are guilty of academic misconduct, not the ghostwriter.

Despite some initial research on determining a particular person's writing style, it is not currently possible to determine whether a text has been written by a certain person. The only recourse educators have is to observe the process of writing, for example, by looking at draft versions of the papers. This will, unfortunately, increase the time and effort that will have to be invested in reading, grading, and giving feedback.

However, since there are students who just hand in the digital version that they purchased from the ghostwriter, a quick look at the properties menu in the Word document or in the PDF can sometimes reveal interesting information. If the author is listed as someone entirely different from the student, or is completely clean, as will be the case for professional ghostwriters, the student can be asked to come to office hours to explain the discrepancy and also to discuss the paper topic with the professor. When simple questions such as where they obtained obscure material or why they chose this approach over another one cannot be answered, it can be an indication that the paper was not written by the student.

2.3.2 Contract Cheating

Robert Clarke and Thomas Lancaster (2006, p. 2) identified a specialized type of ghostwriting they call *contract cheating*, "the process of offering the process of completing an assignment for a student out to tender." There are specialized sites where students can upload their homework assignments or the topics of papers they are supposed to write. Potential authors then bid on completing the assignment, with the student in general choosing the lowest bidder. For many people in underdeveloped countries, even if the pay is low for such a service, it is quite worth their while to take on such jobs. Payment is often handled by services such as PayPal. One of the largest such agencies is freelancer.com (formerly known as vWorker and before that Rent-A-Coder). In order to detect such illicit help, a teacher needs to be attuned to unusual language, and to regularly ask students questions in person about the results of their efforts.

2.3.3 Honorary Authorship

Sometimes so-called *honorary authors* are listed on a paper or a book. There are a number of different variations on this theme and reasons for including such authors, both among students and researchers.

For students who are being asked to prepare an assignment in a team, there are sometimes reasons that students will include the name of another student on the paper or project, even though they did not contribute much to the final effort. Sometimes the students will trade off courses: "I'll do mathematics if you will do the computing assignment." Shelley Yeo (2007, p. 211) has noted that students, especially first-year students, are often surprised to discover that this is not acceptable behavior. Unless one has regular progress report meetings held with a student group, it can be difficult to determine who did the work and who is an "honorary" team member.

With researchers, sometimes a well-known author is included on a paper as an author. These people are not always aware that their name is being used in this manner. The true authors of the paper hope that with the inclusion of a "good name" their chances of being accepted will be increased. Sometimes the principal investigator in a research project insists on being listed as an author on all papers produced in the research group since they obtained the financing, even though they in fact did not do the research or contribute to the writing. This helps them keep their personal citation index at a steady, high level, which seems to be important for obtaining future financing.

The moment problems arise, however, whether the paper turns out to be plagiarized or the data to be cooked in some fashion, these people are often quick to note that they were not really the authors. Many journals are now insisting that all authors sign a paper asserting that they were indeed involved in the production of the paper and/or the results, instead of just the principal investigator, although there have been cases in which these signatures have been forged as well. Other journals now insist on each paper including a section that lists the exact contribution of each author.

An interesting variation is the intentionally dropped authorship, as was noted by Birgitta Forsman (2010) in the Swedish online research journal *Forskning och Framskritt* about a paper that appeared in *Lancet*. The inventor of a technology that was to be studied was co-owner of the company that used the technology and was financing the study. The inventor acted as monitor for the study and was responsible for the quality control, although he was not listed as an author on the paper. This makes the research appear to be independent, when it actually was sponsored by the company in question.

The rule for authorship is rather quite simple: Only those who did the research and wrote the paper should be on it, and all of those who did the research and contributed to writing the paper should be on it. Only obtaining financing or proof-reading a manuscript does not establish authorship. But substantial contributions do, and must be mentioned.

2.3.4 Falsifying Data

Falsifying scientific data is nothing new and certainly not something that has just arisen with the advent of computers. The English inventor Charles Babbage defined a number of methods of falsifying data in the 19th century, noting that (Babbage 1830, pp. 174–175)

> [t]here are several species of impositions that have been practised in science, which are but little known, except to the initiated, and which it may perhaps be possible to render quite intelligible to ordinary understandings. These may be classed under the heads of hoaxing, forging, trimming, and cooking.

Babbage continues with extensive definitions of these terms and gives some contemporary examples.

Hoaxes are quite elaborate falsifications that involve producing an artifact that is said to be an exciting discovery. Heinrich Zankl discusses many hoaxes at length in his book about false science (Zankl 2003). For example, he notes that Arthur Smith Woodward and Charles Dawson presented a skull to the Geological Society in London in 1912 that they purportedly had found near the village of Piltdown. This skull was the supposed missing link between apes and men, about 500,000 years old. The two kept finding more skulls in the area, so it seemed to be true. It was not until 1935 that a geologist proved that the skulls were in a wrong strata for their supposed age, and a few years more until chemical dating proved that the skulls were only 50,000 years old at the most. Another elaborate hoax Zankl discusses was the discovery in 1965 of the so-called "Vineland Map" that was supposed to prove that the Vikings had discovered America. Unfortunately, the ink turned out to be from around 1920, although some researchers are still of the opinion that the map might actually have originated with the Vikings and been redrawn.

Forging data involves making them up. In the 1970s, the English psychologist Cyril Burt was accused of fabricating the data for his extensive studies of twins in order to demonstrate that intelligence is inherited. There are quite a number of discussions published about the fabrications. A good overview is given by Plucker (2013). Burt apparently reused the same correlation coefficient in quite a number of his papers, as was published on the front page of *The Times* on 27 October 1976[1] by the medical correspondent Dr. Oliver Gillie (1977):

> The most sensational charge of scientific fraud in this century is being leveled against the late Sir Cyril Burt, father of British educational psychology. Leading scientists are convinced that Burt published false data and invented crucial facts to support his controversial theory that intelligence is largely inherited.

Trimming data involves discarding outlier data that does not fit the hypothesis. *Cooking data* is the process of making many measurements and then only reporting those choice measurements that are deemed satisfactory by the appropriate standards.

Students will also be falsifying data when they copy measurements done by other students into their lab books instead of doing the work themselves. Lab reports are sometimes handed down from year to year, with minor adjustments made so that the data looks more or less realistic.

2.4 A German Standpoint on Academic Misconduct

In 1998 the German research financing organization Deutsche Forschungsgemeinschaft[2] published sixteen recommendations for good academic practice (Deutsche Forschungsgemeinschaft 1998, Appendix C). This was a result of a case of widespread data falsification scandals involving Friedhelm Herrmann and Marion Brach (Deutsche Forschungsgemeinschaft 2000; Finetti & Himmelrath 1999).

In the final report, these cancer researchers were found to have falsified data in 94 out of 347 publications, and in a further 121 publications there were suspicions of falsified data that could neither be confirmed nor rejected (Tuffs 2000). There were also questions of plagiarism and honorary authorship that could not be clarified by the investigating committee. As a result, both Herrmann and Brach were terminated as professors by their universities, although Hermann continues to work in cancer research as a private doctor.

Reading through the recommendations it is clear that none of the aspects of academic misconduct discussed above are to be tolerated. Most particularly, recommendation 8 states that each university is to set up an ombud for good academic practice, a person or group of persons who are responsible for dealing with cases of academic misconduct that arise at the university.

The DFG itself had an ombud for good academic practice called the *Ombudsman für die Wissenschaft* that is now charged with acting as a general ombud for research in Germany and not only for research funded by the DFG. This body consists of three German professors who investigate complaints. The current members

---

[1] Plucker gives the date as 24 October 1976, but a reprint of the article at (Gillie 1977) gives 27 October 1976 as the date. The online archive of *The Times* does not go back this far.

[2] The *Deutsche Forschungsgemeinschaft* (German Research Foundation) is a self-governing organization for funding science and research in Germany.

(2013) are Wolfgang Löwer, Brigitte M. Jockusch, and Katharina Al-Shamery. They have, however, no legal sanctions that can be leveled, should it be found that a complaint is justified. The DFG recommendations also make it clear that it is the job of the universities to teach their students and young researchers about good academic practice. However, the author's personal experience when addressing the topic with an academic audience is that some researchers are still rather unaware that such a set of recommendations even exists.

The DFG rules are quite important for researchers in Germany, as many rely on funding through this body for their research programs. In 2010 the DFG also instituted a new practice called "Quality not Quantity" (Deutsche Forschungsgemeinschaft 2010) in order to try and stem the tide of multiple publication of the same material. New applications for funding from the DFG are only permitted five publications per applicant CV. This forces the applicant to choose the five best publications, and it minimizes the attraction of non-refereed, pay-to-publish offers. In addition, when submitting the final report on the use of the funds, the report is limited to just two publications per year of financing. They will also only accept papers that have been published, not any that are in preparation, or only submitted at the time of the report.

2.5 Incidence of Plagiarism

A question that is often asked by journalists is: Has the number of students who plagiarize increased since the Internet came into being? This is a question that cannot be answered, since we really do not know how many students plagiarize. We can only know how many we have discovered – not how many have actually plagiarized.

There have been a few investigations that have asked students to self-report how often they have cheated, which includes not only plagiarism but also cheating on tests and other sorts of student academic misconduct. One of the earliest investigations in the UK was conducted by Franklyn-Stokes & Newstead (1995). They observed that cheating was perhaps more prevalent in British universities than staff were aware of. They noted that students were not being properly informed about unacceptable behavior and that they felt that, as mere students, their work did not really count. With the increasing number of students, the type of assessment used is shifting to those kinds of coursework that can more easily be copied or colluded on. Their study suggests that the increased focus on such types of assessment is encouraging cheating behavior in students.

In Germany, in a 2013 study by Sattler, Graeff, & Willen (2013), the authors found 79% of students questioned ($n = 2806$) self-reporting that they had in some way cheated during the past six months. 94% of those who had cheated said that they were never caught. This is exactly the point that Franklyn-Stokes & Newstead were trying to make – cheating is going on, full stop. What is the university doing in order to put a stop to it? It does not really matter if only 10% or 30% or 80% of the students are cheating; the magnitude of the problem should not affect the decision of a university to take action. Cheating and plagiarism happens, and it is bad because people who engage in such behavior are not properly prepared for the workplace or for continuing on in academia.

2.6 Reasons for Plagiarizing

There have been studies published about student cheating since at least the 1940s. Davis, Grover, Becker, & McGregor (1992) gave a brief overview of these works, suggesting that students cheat because of stress and in order to get good grades. They themselves sampled over 6,000 students in both high schools and colleges. Although the vast majority of students (over 90%) say that it is wrong to cheat, they find that over three-quarters of the students have self-reported as having cheated, although the numbers go down as the students progress from high school to college.

They found males more likely to cheat than females, and confirmed stress and the pressure to get good grades as the main reasons for cheating. They also found that cheating was seldom discovered, and even when it was, there were very seldom consequences to be feared, confirming earlier reports. It also seemed that students who were more able to deal with the work were less likely to cheat.

Lambert, Hogan, & Barton (2003) gave a good overview of some of the extensive body of literature on how many students plagiarize and on the reasons given for plagiarizing, as well as reporting on an investigation they conducted at one university in the USA. They found studies giving alienation and low levels of commitment as reasons for cheating, as well as having higher loyalties to friends than to academic standards. Other studies reported the need for good grades as a justification given or found students feeling that cheating was okay because the professor was felt to be unfair. It would go far beyond the scope of this book to address all of these studies in this volume.

An important question is what exactly is meant by the term *cheating*. Franklyn-Stokes & Newstead (1995, p. 170) noted when comparing results from North America with the UK, the former has had to cope with large numbers of students and thus has introduced multiple choice tests as assessments of learning. When authors from the USA or Canada write about cheating, they often only mean cheating in a test-taking situation. In their studies, however, Franklyn-Stokes & Newstead have looked at a much broader definition of cheating. They have a long catalog of student behavior that can be considered to be cheating. Appendix B lists the items from this catalog.

One of the studies Franklyn-Stokes & Newstead report on (1995) also looked into the reasons why students cheat. They confirmed that the principal reasons were time pressure and a desire to get a better grade. They also asked something that other researchers had not asked: What would keep people from cheating? It turned out that fear of punishment is *not* a determent. Rather, only when the students felt that cheating was unnecessary, or when they felt strongly that to cheat would be dishonest, would they refrain from the behavior listed. This means that just identifying and punishing cheaters will not generally be successful. Instead, there must be a focus on educating the students as to what is considered to be cheating behavior and instilling in them a sense of "we don't cheat here."

This has been confirmed more recently by Donald L. McCabe's continuing investigations, as he has reported in numerous papers and talks given worldwide (see, for example, Hughes & McCabe 2006a; McCabe & Pavela 2000; McCabe, Treviño, & Butterfield 2002).

In Germany, Sebastian Sattler from the University of Bielefeld led a study called *FairUse* that tried to determine the factors that contribute to plagiarism. The study uses Rational Choice Theory to postulate that students weigh the costs of plagiarism (doing the research and investing the time to write vs. just finding something suitable to plagiarize) against the benefits (saving time and getting good grades at the same time vs. the chance of being caught and the severity of the probable punishment).

The findings by Sattler et al. (2013) include evidence that the higher the expected utility derived from plagiarism, the more often students plagiarize. The psychological stress that results from breaking rules (internalized norms) seems to reduce the number of incidents of plagiarism, however, so working to instill these norms in students should be helpful in preventing plagiarism. The more chances there are of plagiarizing, however, the more plagiarism there is. But students do seem to spend time deliberating about whether to plagiarize or not, they note, and don't just plagiarize as a habit.

A study by Margaret Platt Jendrek reported that 74% of the students interviewed ($n = 776$) indicated that they had witnessed cheating (Jendrek 1992, p. 262), but less than 1% of those who observed the cheating had actually reported it to their instructor, although there were institutional policies in place requiring them to do so (1992, p. 263). Jendrek notes that she had also investigated faculty reactions to academic misconduct in 1989. That study found that most faculty did not act on cases in which students were caught cheating, often because faculty were not familiar with the processes for dealing with such cases.

Hughes & McCabe (2006a, p. 56) summarize the literature on faculty reluctance for dealing with academic misconduct by noting that faculty often do not even accept the formal policies and procedures of the institution, which they find too complicated and the penalties not appropriate to the misconduct, or that there is no institutional backing for pursuing such cases. Other factors, as mentioned by Hughes & McCabe, included the massive time investment and effort needed for bringing forward such a case that brings no immediate benefit for the faculty. There is also a discomfort factor involving the need to confront a student, coupled with a fear of legal proceedings, or of harassment from students, or even of being blamed by the administration for causing trouble. It is more comfortable for all concerned to just look the other way.

2.7 Why Is Plagiarism a Problem?

The previous sections on plagiarism and academic misconduct have demonstrated the wide range of ethically problematic behavior that is often only called plagiarism.

Sometimes the question arises as to why plagiarism or academic misconduct is a problem at all for schools and universities. An author does not suffer material loss just because a student is using his or her words. Why should teachers get all worked up about that? And why should students be angry when their teachers publish the students' writing as their own – they did participate in the research, and they organized the funding. So should their names not be prominent on the paper? And in the case of plagiarism in dissertations, some universities in Germany have argued that the "missing" quotation marks were just technical weaknesses in the thesis, but that the "main research" was still solid.

First it needs to be clear that when *academics* plagiarize they are damaging academic discourse. They are obscuring facts, for example, who came up with which idea, and that makes it harder for those coming after to discover what the truth actually is. This is also a problem when data is fabricated or when assertions are just copied without checking if they are, indeed, true. Such data will first be accepted by other researchers, who might then waste time and resources trying to replicate the results. Or they will be building on top of data that is not accurate.

Plagiarisms also create redundancy when a reader discovers that he or she is now reading something that they have already read before. This is actually the source of many plagiarism and multiple publication discoveries: Young researchers are trying to obtain an overview of an area and stumble upon the copies. Duplicate publication and plagiarism make academic processes take much longer, as one needs to perhaps wade through 100 papers instead of 10 in order to find what one is looking for and who the original author is. There is also damage sustained by the journals if they repeatedly publish fabricated or plagiarized papers. That is why many important academic journals take retractions very seriously. Journals then need to investigate: What went wrong in the peer review process that is supposed to weed out bad science? Scientists will wonder how many other articles are lurking in this journal that are also plagiarisms or statistically flawed or otherwise a result of academic misconduct. What can be believed? The integrity of the journal is at stake.

When researchers publish multiple copies of the same research or plagiarizes another's papers, they are at the same time inflating their own CVs and making their worth as a researcher appear to be more productive than is actually the case. This gives the researcher an unfair advantage in situations where candidates for jobs are ranked by the number of publications, or when a grant is given to the researcher with the most publications or the best citation count.

The second problem concerns plagiarizing *students*. People are at school and university in order to learn. They are to learn how to do research, how to structure the results, and to present them in their own words, so that others can profit from the research. If they use the words or research of others, it is necessary that this be clearly marked. If a student submits a plagiarism, he or she loses out on a learning experience, even if it is just a paper that will not be published. Certainly, students may learn something by compiling what others have said about a topic. But if they do not start learning how to excerpt and quote as early as possible, it will be that much harder for them to learn how academic writing is done later on.

In addition, the grade that is given for an assignment is usually part of a final grade that is a part of a certification of some sort. This is the basis for a degree, and especially in Germany, it is often more important for someone to have a proper degree than for them to have the skills that are necessary for a job, especially for higher-paying jobs. It is unfair to other students if plagiarism brings rewards, similar to doping in sports, as long as it remains undiscovered. Indeed, not all graduates will go on to be scientists, but it is important that college students learn how to acquire large amounts of information and to structure it so that they can find it quickly. Even if they will only be using information and not generating new knowledge, it is important that they understand how to give credit for the work of others.

A third issue has to do with *professors* assuming authorship for work done by students. Time and again the author has heard that some university professors feel that they are owed a sort of *jus primae noctis*, a right of first use, so to speak, from researchers who are working on projects for which the professors have obtained the financing. This attitude seems to be especially prevalent in the fields of medicine and biochemistry. By all means, the principal investigator perhaps came up with the research idea and organized the money for the research. Obviously, this should be noted in the published results. But the only persons who should be listed as authors are those who did the research and the writing. And it should be clear that all authors take responsibility for the text. They cannot step back and point fingers if it is later determined that there are problems with the manuscript.

It is troubling that, especially in Germany, talented young academics have a difficult time being integrated into the research process. They are not permitted to lead a research group; this honor goes instead to the senior professor who sometimes does not actually participate in the research itself. Perhaps putting a stop to this practice would be beneficial in making it clearer that the young researchers are the ones doing the work.

Professors who publish the results of their students or even obtain patents based on the work that others have done should be aware of the fact that this is academic misconduct. It is rather simple to see why this is so: If the students produced the research and/or the writing for a grade, then it must only be the work of the student and thus the professor has no authorship claim. If, however, the professor has enough invested in the research to be considered an author, then the material cannot be used for obtaining a grade.

Finally, in the case of *plagiarized dissertations*, a candidate now has a title that is included on their identity cards in Germany, and tends to be engraved on a brass plate on the door at home and work, as well as now possessing the basic qualification for teaching at an engineering college. They are also eligible for postdoc positions, taking financing away from those who would perhaps be better qualified for the position. Or they will be promoted to higher levels of responsibility within the company for which they are working. Once too many people who have technical weaknesses in their research are ingrained in such institutions, the research climate will suffer and it will be difficult for them to teach students how to be scrupulously honest about what parts of a paper are from whom.

All of these aspects are ethically problematic and can even raise legal issues, which cannot be dealt with in this book. But it is our duty as teachers to show students how to do research and to teach them to write, no matter what field we are working in. We must insist that they write in their own words, and that things stated as facts are, indeed, facts.

It is important that students are given a sense of why exactly they are studying. There seems to be a tendency to focus just on grades, even feelings of entitlement to good grades just for enrolling in a course, and to concentrate only on obtaining credits. The notion that one attends university to learn something, to satisfy one's curiosity, or to learn to think, seems to be quite old-fashioned today. This is aggravated by employers focusing on having diplomas and good grades before candidates are even looked at, or as a means of sorting out applicants. Even though it is understandable when there are many applicants for a position, it should not be the only aspect that is considered.

References

Babbage, C. (1830). *Reflections on the Decline of Science in England: And on Some of its Causes*. London: B. Fellowes. Available as a digitalized facsimilie at http://reader.digitale-sammlungen.de/de/fs1/object/display/bsb10730620_00005.html and as a text extract at http://www.gutenberg.org/files/1216/1216-h/1216-h.htm cited 1 September 2013.

Clarke, R. & Lancaster, T. (2006). Eliminating The Successor To Plagiarism? Identifying The Usage Of Contract Cheating Sites. In: *Second International Plagiarism Conference: Prevention, Practice and Policy*, Newcastle, UK, 19–21 June. Available at http://archive.plagiarismadvice.org/documents/papers/2006Papers05.pdf cited 12 March 2013.

Davis, S. F., Grover, C. A., Becker, A. H., & McGregor, L. N. (1992). Academic dishonesty: prevalence, determinants, techniques and punishments. In: *Teaching of Psychology*, Vol. 19, No. 1, pp. 16–20.

Deutsche Forschungsgemeinschaft. (1998). *Vorschläge zur Sicherung guter wissenschaftlicher Praxis. Empfehlungen der Kommission „Selbstkontrolle in der Wissenschaft". Denkschrift*. Weinheim: Wiley-VCH. Available at http://www.dfg.de/download/pdf/dfg_im_profil/reden_stellungnahmen/download/empfehlung_wiss_praxis_0198.pdf cited 5 July 2013.

Deutsche Forschungsgemeinschaft. (2000). *Task Force legt Abschlußbericht vor*. [Press release], No. 26, 19 June. Available at http://www.dfg.de/service/presse/pressemitteilungen/2000/pressemitteilung_nr_26/index.html cited 23 July 2012.

Deutsche Forschungsgemeinschaft. (2010). *„Qualität statt Quantität" – DFG setzt Regeln gegen Publikationsflut in der Wissenschaft*. [Press release], No. 7, 23 February. Available at http://www.dfg.de/service/presse/pressemitteilungen/2010/ pressemitteilung_nr_07/index.html cited 3 January 2013.

Dreier, T. & Ohly, A. (2013). *Plagiate: Wissenschaftsethik und Recht*. Tübingen: Mohr Siebeck.

Englisch, P. (1933). *Meister des Plagiats oder die Kunst der Abschriftstellerei*. Berlin-Karlshorst: Hannibal-Verlag. Digitized version available at http: //visuallibrary.net/ihd/content/titleinfo/122128 cited 22 August 2013.

Finetti, M. & Himmelrath, A. (1999). *Der Sündenfall: Betrug und Fälschung in der deutschen Wissenschaft*. Stuttgart: J. Raabe Verlag. The chapter on the Herrmann/Brach case is available online at http://www.wiwo.de/technologie/ forschung/auszug-der-suendenfall-der-fall-herrmann-brach/8280516.html cited 13 August 2013.

Fishman, T. (2009). "We know it when we see it" is not Good Enough: Toward a Standard Definition of Plagiarism that Transcends Theft, Fraud, and Copyright. In: *Proceedings of the Fourth Asia Pacific Conference on Educational Integrity* (4APCEI) 28–30 September, University of Wollongong, NSW, Australia. Preprint available online at http://www.bmartin.cc/pubs/09-4apcei/4apcei-Fishman.pdf cited 12 March 2013.

Forsman, B. (2010). Oredlighet i forskning. In: *Forskning & Framsteg*, 27 September. Available at http://www.fof.se/blogg/birgitta-forsman/oredlighet-i-forskning cited 12 March 2013.

Franklyn-Stokes, A. & Newstead, S. E. (1995). Undergraduate Cheating: Who does what and why? In: *Studies in Higher Education*, Vol. 20, No. 2, pp. 159–172.

Gibaldi, J. (1998). *MLA Style Manual and Guide to Scholarly Publishing*. (2nd ed.) New York: Modern Language Association of America.

Gibaldi, J. (2003). *MLA Handbook for Writers of Research Papers*. (6th ed.) New York: Modern Language Association of America.

Gillie, O. (1977). Did Sir Cyril Burt Fake His Research on Heritability of Intelligence? Part I. In: *The Phi Delta Kappan, Technology and Education*, Vol. 58, No. 6, February, pp. 469–471. Available at http://www.jstor.org/stable/20298643 as a reprint of *London Sunday Times*, 27 October 1976. cited 30 July 2013.

Gumm, H. P. (2010). *Plagiarism or „naturally given"? Decide for yourself...* [Web site]. http://www.mathematik.uni-marburg.de/~gumm/Plagiarism/ cited 4 August 2013.

Guttenberg, K.-T., Freiherr von und zu. (2009). *Verfassung und Verfassungsvertrag: Konstitutionelle Entwicklungsstufen in den USA und der EU*. Berlin: Duncker & Humblot.

Haller, G. (2003) *Recht – Demokratie – Politik: Zum unterschiedlichen Verständnis von Staat und Nation dies- und jenseits des Atlantiks*. Talk given at the conference «Die USA – Innenansichten einer Weltmacht», 7–8 February, Catholic Academy in Bavaria, Munich. Transcript available at http://www.grethaller.ch/texte/2003/ recht-demokratie-politik_europa-und-usa.html cited 26 July 2013.

Harvard University (2013) *Harvard Guide to Using Sources: Avoiding Plagiarism – What Constitutes Plagiarism?* http://isites.harvard.edu/icb/icb.do?keyword= k70847&pageid=icb.page342054 cited 2 August 2013.

Howard, R. M. (1999) *Standing in the Shadow of Giants: Plagiarists, Authors, Collaborators*. Stamford, CT: Ablex Pub.

Hughes, J. M. C. & McCabe, D. L. (2006a) Understanding Academic Misconduct. In: *The Canadian Journal of Higher Education/La revue canadienne d'enseignement supérieur*, Vol. 36, No. 1, pp. 49–63.

Jendrek, M. P. (1992) Students' Reactions to Academic Dishonesty. In: *Journal of college student development*, Vol. 33, No. 3, pp. 260–273.

Kühnhardt, L. (2004) Auf dem Weg zu einem europäischen Verfassungspatriotismus. In: *Neue Zürcher Zeitung*, Vol. 225, No. 163, 16 July, p. 9.

Lahusen, B. (2006) Goldene Zeiten: Anmerkungen zu Hans-Peter Schwintowski, Juristische Methodenlehre, UTB basics Recht und Wirtschaft 2005. In: *Kritische Justiz*, Vol. 39, No. 4, pp. 398–417. Available at http://www.kj.nomos.de/fileadmin/kj/doc/2006/20064Lahusen_S_398.pdf cited 12 August 2013.

Lambert, E. G., Hogan, N. L, & Barton, S. M. (2003) Collegiate Academic Dishonesty Revisited: What Have They Done, How Often Have They Done It, Who Does It, And Why Did They Do It? In: *Electronic Journal of Sociology*, Vol. 7, No. 4. Available at http://www.sociology.org/content/vol7.4/lambert_etal.html cited 12 March 2013.

Lancaster, T. (2003) *Effective and Efficient Plagiarism Detection*. [PhD thesis], London South Bank University, London, UK. Available at http://bcu.academia.edu/ThomasLancaster/Papers/79311/Effective_and_Efficient_Plagiarism_Detection cited 12 March 2013.

lol. (2008). How many words do you have to change in a document for it not to be plagiarised? [Online Forum Question]. http://answers.yahoo.com/question/index?qid=20080504003338AAegQeG cited 10 July 2013.

Maurer, H., Kappe, F., & Zaka, B. (2006). Plagiarism – A Survey. In: *Journal of Universal Computer Science*, Vol. 12, No. 8, pp. 1050–1084. Available at http://citeseerx.ist.psu.edu/viewdoc/download?doi=10.1.1.102.5629&rep=rep1&type=pdf cited 30 June 2013.

McCabe, D. L. & Pavela, G.R. (2000). Some Good News about Academic Integrity. In: *Change*. Vol. 32, No. 5, pp. 32–38.

McCabe, D. L., Treviño, L. K., & Butterfield, K. D. (2002). Honor codes and other contextual influences on academic integrity: A replication and extension to modified honor code settings. In: *Research in Higher Education*, Vol. 43, No. 3, pp. 357–378.

McIver, S. B. (1994). *Dreamers, Schemers and Scalawags*. Sarasota, FL: Pineapple Press.

Modern Language Association. (2008). *MLA Style Manual and Guide to Scholarly Publishing*. (3rd ed.) New York: Modern Language Association of America.

Münch, I. von (2012). *Gute Wissenschaft*. Berlin: Duncker & Humblot.

O'Toole, G. (2010). If You Steal From One Author, It's Plagiarism; If You Steal From Many, It's Research. In: *Quote Investigator*. [Blog], 20 September. http://quoteinvestigator.com/2010/09/20/plagiarism/ cited 5 August 2013.

Plucker, J. (2013). Cyril L. Burt. In: *Human Intelligence*. [Web page] http://www.intelltheory.com/burt.shtml cited 2 July 2013.

Roig, M. (2003). *Avoiding plagiarism, self-plagiarism, and other questionable writing practices: A guide to ethical writing.* Revision from 2006 available at http://www.cse.msu.edu/~alexliu/plagiarism.pdf cited 14 August 2013.

Sattler, S., Graeff, P., & Willen, S. (2013). Explaining the Decision to Plagiarize: An Empirical Test of the Interplay Between Rationality, Norms, and Opportunity. In: *Deviant Behavior*, Vol. 34, No. 6, pp. 444–463. Available at http://dx.doi.org/10.1080/01639625.2012.735909 cited 14 August 2013.

Scanlon, P. M. (2007). Song From Myself: An Anatomy of Self-Plagiarism. In: *Plagiary: Cross-Disciplinary Studies in Plagiarism, Fabrication and Falsification*, Vol. 2, pp. 57–66. Available at http://hdl.handle.net/2027/spo.5240451.0002.007 cited 14 August 2013.

Stein, B. & Meyer zu Eissen, S. (2007). Intrinsic Plagiarism Analysis with Meta Learning. In: B. Stein, M. Koppel, & E. Stamatatos (Eds.), *Proceedings of the SIGIR Workshop on Plagiarism Analysis, Authorship Identification, and Near Duplicate Detection (PAN 07)*, Amsterdam, Netherlands, 27 July, pp. 45–50. Available at http://ceur-ws.org/Vol-276/ cited 15 July 2013.

Tuffs, A. (2000). Fraud investigation concludes that self regulation has failed. In: *British Medical Journal*, Vol. 321, No. 7253, 8 July, p. 72. Print version available at https://www.ncbi.nlm.nih.gov/pmc/articles/PMC1127761/ and a longer article available online at http://www.ncbi.nlm.nih.gov/pmc/articles/PMC1173372/

Weber-Wulff, D. & Wohnsdorf, G. (2006). Strategien der Plagiatsbekämpfung. In: *Information: Wissenschaft & Praxis*, Vol. 57, No. 2, pp. 90–98.

Yeo, S. (2007). First-year university science and engineering students' understanding of plagiarism. In: *Higher Education Research & Development*, Vol. 26, No. 2, pp. 199–216.

Zankl, H. (2003). *Fälscher, Schwindler, Scharlatane: Betrug in Forschung und Wissenschaft*. Weinheim: Wiley-VCH.

Chapter 3
Plagiarism in Germany

This chapter discusses the special situation in Germany with respect to the public plagiarism documentation done in the GuttenPlag and VroniPlag Wikis, as well as discussing the reactions of the universities to the uncomfortable revelations. Furthermore, Germany has a long tradition of plagiarism and plagiarism accusations, as will be demonstrated by a selection of historical cases of plagiarism. The chapter closes with a discussion of how dissertations are prepared in Germany today as this does differ from the situation in other countries.

3.1 zu Guttenberg and the Crowd

The year 2011 was the year that Germany finally became acutely aware that it, too, has a major plagiarism problem. In February of that year, the doctoral dissertation in law of the popular Minister of Defense, Karl-Theodor Freiherr von und zu Guttenberg (born 1971), that had just recently been published (zu Guttenberg 2009), was reviewed by Andreas Fischer-Lescano, a law professor in Bremen (Fischer-Lescano 2011).

In 2007 zu Guttenberg was awarded the top German grade for a doctorate, *summa cum laude*, by the University of Bayreuth. His thesis dealt with the stages of development of constitutions in the USA and in the European Union and questions such as the use of religious phrases in the constitution and was subsequently published by one of the top legal publishing houses in Germany. Fischer-Lescano found some text parallels and published them – he informed the press at the same time. Roland Preuß and Tanjev Schultz, journalists with the daily newspaper *Süddeutsche Zeitung*, quickly came out with a page two article about the text closeness on February 16, 2011 (Preuß & Schultz 2011a)[1]. The German press went into a frenzy, publishing thousands of articles in the coming days.

---

[1] The online version of this article in the *Süddeutsche Zeitung* (Preuß & Schultz 2011b) has a different but similar text. The two journalists also wrote a book together (Preuß & Schultz 2011c), detailing their perception of what happened during those two extremely intensive weeks. Karl-

Germany was again a divided country – this time not geographically, but politically. Many supporters of the charismatic law & order politician zu Guttenberg, a member of the conservative Bavarian CSU party, were angry that he was being accused of what they felt was such a trivial offense. Surely everyone had cheated in school? On the other side of the divide, intellectuals and academics were outraged that the basis of academic inquiry, honesty, was being belittled by zu Guttenberg's defenders. There were Facebook fan groups for zu Guttenberg and petitions from professors and doctoral students demanding that he step down.

And then there was a group of people who were irritated at zu Guttenberg's attempts to play the affair down. He had called the notion that he had plagiarized "abstruse" (Spiegel Online 2011). Some people felt that, in the face of at least nine pages of word-for-word plagiarism, what was abstruse was his denial of the evidence. They decided to document the plagiarism, assuming rightly that where there was some plagiarism of this quality, there would be more. A wiki was soon set up called the GuttenPlag Wiki (2011a), a web space for collaborative writing, and it was used for publicly and collaboratively documenting the extent of the plagiarism, as one of the founders of the GuttenPlag Wiki, PlagDoc, writes (PlagDoc & Kotynek 2012).

With the press reporting constantly about the scandal and the documentation effort, many more people joined in. And since it was so simple to successfully find plagiarism – zu Guttenberg had used numerous sources available online – the amount of documented plagiarism grew rapidly. People were stepping over each other's toes, duplicating work. A small group of wiki administrators stepped up and began coordinating the work. Rules were developed to make sure that people were not stating plagiarism where there was none so that a workflow ensued. And the idea arose to visualize the parts of the thesis that still needed to be looked at using a sort of barcode.

The press jumped on this visual cue as the people documenting the plagiarism remained pseudonymous and were thus unavailable for pictures and home stories on why they were doing this. The barcode (which will be explained in more detail in Sect. 4.4.6) came to be an indicator of the extent of the plagiarism – and it was extensive indeed. When the dust settled, the thesis was found to contain plagiarism on 94% of the pages. Some people went to the trouble of documenting the number of lines of the text that were plagiarized: that turned out to be almost two-thirds of the thesis.

The Bavarian University of Bayreuth swiftly investigated the issue and rescinded zu Guttenberg's doctorate after a mere two weeks' of deliberation. A few months later, in May 2011, the university took an unusual step and released the report of the committee that investigated the thesis (Kommission Bayreuth 2011). Normally such reports remain secret. But since this was the law department of the university, they took care to collect up the relevant legal rulings about plagiarism and to include them into their considerations. That makes this document a very valuable resource, in addition to being an excellent documentation of the reasons why the dissertation

---

Theodor zu Guttenberg himself was interviewed by journalist Giovanni di Lorenzo (zu Guttenberg & di Lorenzo 2011) about his personal view of the situation.

was rescinded. The university gave a press conference to report their decision, with both the university president and the dean giving statements.

There was much shaking of heads about how it was possible that the advisor and the grader could have missed seeing such obvious plagiarisms, even plagiarisms from their own works, from standard texts on the topic, and from quite a number of newspaper articles. The list of sources can be found on the GuttenPlag Wiki (2011b). Even the first page was plagiarized – from an article by Barbara Zehnpfennig in the *Frankfurter Allgemeine Zeitung* about the USA that begins: "'E pluribus unum', out of many, one" (Zehnpfennig 1997). Pundits noted that this was, fittingly, a perfect motto for the thesis.

Peter Häberle, the doctoral advisor, is a well-respected law professor who stated that he just could not have imagined that zu Guttenberg would have done something like this, as he was one of his best students, but that the shortcomings of the thesis were inconceivable (Häberle 2011). Many academics reassured themselves that this issue was most certainly a singularity. He was surely just one charming personality who was not happy with only being a baron and a member of parliament, but had also felt the necessity of having a doctoral degree – which in Germany implies having it as part of one's name and being called *Herr Doktor* or *Frau Doktor* in everyday life. As it has turned out, this was not just an isolated case. Plagiarism was and is widespread, as in particular the VroniPlag Wiki has shown.

And even when it is discovered and reported, more often than not the situation is played down or even covered up. The journalist Manuel Bewarder (2011) reports on a doctoral student at the University of Münster who was working on his doctorate in law during the summer of 2010. He had obtained the thesis of zu Guttenberg and some of the related works and quickly discovered some plagiarism on the part of zu Guttenberg. The student wrote a paper about the plagiarism and showed it to a few people. He was advised to not publish it, as it would stir up too much trouble and he would find himself in the middle of it all. Only one professor recommended going public with the documentation, but the doctoral student then decided against publication. The doctoral advisor suggested informing the ethics committee of the University of Bayreuth, but did not follow up on the case. And so it took a few more months until Fischer-Lescano was reviewing the book for the extent of the plagiarism to be known and for action to be taken.

3.2 VroniPlag Wiki

A month after zu Guttenberg's doctorate was rescinded, on 28 March 2011, another plagiarism in a law dissertation started to be publicly documented[2]. Some members of the GuttenPlag Wiki group had found plagiarism in the thesis submitted by the daughter of a Bavarian politician to the University of Konstanz in 2008. They set up the documentation operation on a similar wiki that was called the Vroni-

---

[2] The following section is based on the author's own experiences with the group. An overview with links to many important pages can be found on the user page of Plagin Hood (2013).

Plag Wiki, because *Vroni* is the woman's nickname. Again, there was much public agitation about why people were spending all this time documenting plagiarisms in doctoral theses. Critics questioned whether there were political or monetary motives involved, and who these people were who were preparing the documentation.

At the time of writing this book, the home page of VroniPlag Wiki (2013a) lists 48 cases by name, documenting plagiarism in doctoral dissertations, habilitations, a published master's thesis, and a textbook. An overview of all of the cases can be found at the site (VroniPlag Wiki 2013d). Despite countless accusations that VPW targets politicians on the right or center of the political spectrum, a majority of the documented cases are not politicians. Only about a quarter of the cases documented involve someone who is or was politically active, according to the site statistics. (VroniPlag Wiki 2013e). More than a dozen cases, and this is far more troubling, can be seen in these statistics to be by persons who are now professors or honorary professors or university lecturers. If these people could not write a scholarly text without plagiarizing, how can they be teaching students not to plagiarize?

A criticism that is sometimes heard about the activists with the platform is that they are anonymous. Actually, they are pseudonymous, as they use the same *nick* (online name) for all of their edits and comments. There are various reasons for this that have been documented in the FAQ (VroniPlag Wiki 2013b):

> Innerhalb des Wikis soll keine Rolle spielen, wer etwas sagt, sondern nur, was gesagt wird. Die Bekanntheit persönlicher Hintergründe (z. B. Beruf, akademischer Grad) könnte zu Bevormundung und Kompetenzgerangel führen. Außerhalb des Wikis kann es passieren, dass Attribute und Meinungen einer einzelnen Person zu sehr auf das Wiki und seine Mitglieder bezogen werden. Einige Wiki-Beitragende wollen sich durch die Anonymität/Pseudonymität vor persönlichen Anfeindungen schützen und eventuelle Nachteile (z. B. im Beruf) vermeiden, die sie ggf. härter treffen könnten als Journalisten oder anerkannte Gutachter, welche sich auch leichter zur Wehr setzen können (z. B. mithilfe einer Rechtsabteilung oder aufgrund ihres ohnehin öffentlichen Wirkens).

> Within this wiki it does not matter who states something, but only what they say. Knowledge of personal backgrounds (for example, job or academic titles) could lead to patronizing behavior or fights over authority. Outside of the wiki it is possible that attributes or opinions of individuals are linked too closely with the wiki and its activists. Some wiki contributors want to use the anonymity/pseudonymity to guard themselves from personal attacks and avoid potential disadvantages (for example in their jobs). Journalists or tenured experts would not be hit as hard by recriminations and have other means at their disposal (for example a legal department or their public position) to defend themselves.
> *(translation by the author)*

Some concerns have been raised about the names of the plagiarists being put up on the home page, especially the names of those who are not for some reason well-known. There is an air of putting someone into the stocks in the center of town so that they are on display for public ridicule. In a way, since the VroniPlag Wiki site has a rather high Google Page Rank, this is indeed the case. Anyone searching for this person will probably have a link to the site on the first page of the results. Therefore much care must be taken to only publish names when the amount of text parallels documented is so large that the VroniPlag Wiki contributors think that it needs to be made public.

The reason for making the names and the documentation public is that the dissertations must be published in Germany and are thus part of the body of academic knowledge. It needs to be made clear that the author of this particular thesis passed off the work of others as his or her own. A doctorate, which brings many advantages to the bearer, especially in Germany, needs to be rescinded if the basis for granting it did not follow good academic practice. It would be detrimental for following scientists to be basing their own work on the work of a fraud. And since the fraudulent work was submitted by a person, that person must personally bear the consequences.

Are the doctoral students themselves always at fault? Is it not perhaps at least a partial responsibility of the doctoral advisor or the university or the system that so generously rewards persons bearing doctorates? These are valid points that need discussion, but if one understands the significance of a doctorate in Germany – the first major public academic publication – then it becomes clear that it is, indeed, the responsibility of the doctoral student to ensure that good academic practice has been followed.

The main focus of the wiki work is not, however, on the persons, but on the documented theses. It is interesting to see that examples of all types of plagiarism have been found in the examined dissertations, and even a new type was identified and given a name (see Sect. 2.2.8). The theses have been from many different fields (although legal dissertations make up a good third of the cases with medicine coming in second) and from different universities throughout Germany, as well as one from Denmark (case *Nm*)[3], one completed in Poland by a German citizen (case *Cs*), and one submitted to an Austrian university (case *Rm*).

There are six cases that have been documented on VroniPlag Wiki for which it is known that the respective universities have declared the theses not to be academic misconduct or not sufficient in order to rescind the doctorates, despite extensive documentation of text parallels. Not all of the universities announce the results of their investigation, however, so there may be more.

- The case of *Pes*, who submitted a dissertation to the *Fernuniversität Hagen* in law in 2003. The thesis contains some form of text parallel on 24% of the pages. Many of them were close paraphrases or word-for-word copies of a source that was named in a footnote, but without making clear how much exactly was attributed to this author. This kind of text parallel is called a pawn sacrifice, as was defined in Sect. 2.2.7. The university decided not to rescind the doctorate after three experts stated that the text parallels documented did not provide sufficient grounds for doing so (Bossemeyer 2011). The university did not elaborate any further; in particular they did not explain why some of the more troublesome passages were found not to be plagiarisms.
- The case of *Nk*, who submitted a dissertation in medicine to the *University of Heidelberg* in 2002 and for some unknown reason did not defend it until 2006.

---

[3] For many of the VroniPlag Wiki cases, the name of the person submitting the thesis is unimportant. The abbreviations used on the platform for naming the cases will be used here. In order to see the documentation for a particular case, prepend http://de.vroniplag.wikia.com/wiki/ to the abbreviation in order to obtain the web address. All cases are also linked from (VroniPlag Wiki 2013d).

This thesis contains text parallels on over 75% of the pages, and a full two-thirds of these pages were more than three-quarters copied without acknowledgment, mostly from the habilitation – a second dissertation required of candidates for university professorships in Germany – submitted by her own advisor in 1995. It was found by the university that this was still enough of an original work for a dissertation, as the most important part of a medical thesis was the data and not the text surrounding the data (Streuer 2012). Here, too, the university did not publish any details on why exactly the individual fragments could be considered not to be plagiarism.

- The case of *Ut*, who submitted a habilitation to the *Charité* medical school in Berlin in 2003. There are large portions of three doctoral dissertations from students of *Ut* found verbatim in the habilitation without proper acknowledgment. Two of the doctoral dissertations were submitted before the habilitation. One, however, was submitted just a month after the habilitation, although it was not defended until 2005. If the latter one is not counted on account of the formal reason that it was submitted after the habilitation, then there are still text parallels on 34% of the pages in the thesis. But since the habilitation contains some of the same diagrams as that third dissertation, only numbered differently, while the text refers to the diagram numbers exactly as found in this third doctoral dissertation, one may assume that the text was known and thus copied, bringing the amount of pages with text parallels to 79%. Figure 3.1 gives a graphical overview of the extent of usage of these other works. The green and blue marked pages are from papers published by *Ut* with others, the yellow and orange marked pages are dissertations submitted before the habilitation. The red marking is a dissertation submitted after the habilitation. The Charité has stated that since this case has

					7	8	9	10	11	12	13	14	15	16	17	18	19	20
21	22	23	24	25	26	27	28	29	30	31	32	33	34	35	36	37	38	39	40
41	42	43	44	45	46	47	48	49	50	51	52	53	54	55	56	57	58	59	60
61	62	63	64	65	66	67	68	69	70	71	72	73	74	75	76	77	78	79	80
81	82	83	84	85	86	87	88	89	90	91	92	93	94	95	96	97	98	99	100
101	102	103	104	105	106	107	108	109	110	111	112	113	114	115	116	117	118	119	120
121	122	123	124	125														

Fig. 3.1 Text parallels in the habilitation of *Ut* (VroniPlag Wiki 2011b)

already been dealt with, and since the doctoral dissertation authors were thanked in the habilitation and that large portions of the text were taken from publications that the authors had published together, and the doctoral students permitted the use of their texts in such a manner, there is no academic misconduct here.

This is curious, as it violates the German research council DFG's standards of good academic practice (see Sect. 2.4).

- The case of *Dd*, who submitted a thesis in engineering to the *Brandenburg Technical University of Cottbus* in 1999. This thesis is over 44% a more or less verbatim text parallel from nine sources, among them a documentation that was produced at Dd's company. This company also provides external financing for the university. The Brandenburgische Technische Universität Cottbus (2012) decided that the thesis only contains technical weaknesses, but will not release the expertise that might explain why such large-scale copying without proper acknowledgment is acceptable for a dissertation, while a smaller amount taken from the Wikipedia in a student's paper at the same university results in the student's work being graded as 0 points (Brandenburgische Technische Universität Cottbus 2003).
- In the case of *Dv*, a member of parliament who submitted his dissertation in law to the *University of Würzburg*, the university decided after 17 months of deliberation that it was permissible to use text from "general texts", according to a local newspaper report (Viebig 2013). The newspaper quotes the press secretary of the university as stating:

  > Wenn jemand etwa aus UN-Resolutionen oder aus allgemeinen Texten zitiert, muss er das unserer Auffassung nach nicht unbedingt kenntlich machen.

  > If someone quotes from a UN resolution or from general texts, it is not necessary to make this explicit, in our opinion.
  > *(translation by the author)*

  The degree is not being rescinded.
- In the case of *Jg*, the mayor of a town in the Lausitz region who defended his dissertation two years after he became mayor, the *Technical University of Berlin* decided after 18 months of deliberation that there were only technical weaknesses in the citation structure and thus they were not rescinding his doctorate, but he was asked to re-submit the thesis by the end of July 2013 "identically, but with the quotations correctly given" (translation by the author from Terp 2013). A new version of the thesis was handed in on time and the university is still in the process examining it.

These cases notwithstanding, there are currently nine dissertations that have been definitely rescinded and one textbook on academic writing in law has been withdrawn from the market. Much more problematic is that for many of the other cases, the universities have been investigating the cases for a very long time, sometimes for over two years, although VroniPlag Wiki has already extensively documented text parallels that could be considered plagiarism.

The author contacted the universities in question just before going to press. In some cases, a doctorate has been rescinded, but the case has gone to court and is languishing there. In other cases, the university is still investigating. At the University of Bonn the faculty of Arts and Letters was surprised to hear that in addition to their prominent cases, *Gc* and *Mm*, both of which had their doctorates rescinded, they still had a case open, *Mw*. Apparently, there had been a change of dean and the new dean did not know that investigations were supposed to be ongoing.

Some of the emails written to university presidents are just ignored, others an-swer with vague referrals to the protection of personal information demanding se-crecy. Only one university to date, the Free University in Berlin, answered within a week, naming the case number for future referral. That would be what one would expect to be the standard procedure for dealing with such cases. The troubling as-pect of all this is that the persons in question continue using doctorates in public that are based on texts with much publicly documented possible plagiarism. It is not clear why it should take more than six months to decide a case.

Researchers preparing the documentation at VroniPlag Wiki are not focused on the question of rescinding doctorates, however, but only on documenting text paral-lels that could constitute plagiarism that can be found in quite a number of university theses. Historic cases of plagiarism are also of interest, as will be discussed in Sect. 3.5, as these cases demonstrate that plagiarism in academic writing is nothing new and most certainly not something that has just arisen with the advent of the Internet, as is sometimes postulated in published interviews.

Public interest in the work documented on the wiki is still going strong. The statistics will often show a page view spike when there are articles published in the media about a case documented on the wiki. And with the topic of plagiarism often being mentioned in talk shows or even popular TV shows such as "Tatort" (ARD 2013)[4], discussions about plagiarism have become a part of everyday German discourse, even if it has not been possible to elicit much more than just symbolic actions from the stakeholders in the German university system.

3.3 Reactions of the German Universities

The reluctance of many Germany universities to take swift and transparent action on cases of documented plagiarism or academic misconduct gives the impression that they rather hope that if they do not speak about this for some time, it will blow over and they will not have to do anything unpleasant. It is understandable, as the committees who must judge these cases are volunteers who are doing this work in addition to their normal workload. What has happened was a mistake in the past, perhaps even made by someone who is no longer in the department. There is noth-ing to be gained from cleaning up the mess, although the results can be devastating for the plagiarist and give the university supposedly bad publicity. They can also imagine how they would feel if someone found big problems with their own disser-tations. Many universities apparently do not realize that dealing transparently with such cases actually demonstrates that they really care about good academic practice.

In a way, the situation of a university that is informed of possible academic mis-conduct can be seen as a sort of inverse "prisoner's dilemma." This is a situation in

---

[4] This installment of the popular crime television series was first broadcast on 11 March 2012. The forensic examiner Prof. Dr. Karl-Friedrich Boerne points out to a police officer mistakenly calling his forensic medical assistant "Dr. Alberich": "Die ist genauso wenig promoviert wie der Herr zu Guttenberg." (She has just as much of a doctorate as Mr. zu Guttenberg does – none).

which it would be best for all involved persons to act in a certain way, for example, to not speak about a criminal action, but for each individual it would be better to act differently, for example, to accept a lesser sentence for turning in the others. It would be better for the academic community if these cases were dealt with promptly and transparently in order to remove the possible gain that can be had from academic misconduct. For an individual university or even the committee members, however, the personal best interest appears to be saying nothing. It is a lot of work to investigate a case of academic misconduct, and some universities feel that it somehow tarnishes their reputations for cases to have happened.

This is one reason for injecting public attention into the situation. Media reports can break the prisoner's dilemma, as can oversight bodies or institutional changes dedicated to avoiding plagiarism and academic misconduct in the first place. Such changes can be very difficult to set up and institutionalize, although the University of Mainz is one of a number of universities that is publicly taking action to teach students how to avoid plagiarism (Fittkau 2013). They have understood that academic misconduct happens, but that it is important to take measures to avoid it happening and to have good procedures for dealing with the cases that do occur.

There have been calls from various national bodies in Germany to "do something," although generally for someone else to take action. The suggestions have ranged from the futile – making sure all the students swear an oath that they did the work themselves – to implementing the mandatory use of plagiarism detection systems and ultimately to unworkable ideas such as "outlawing" ghostwriting[5].

The problem, however, is not with the existence of ghostwriters. It is not illegal to write a paper for someone else. The problem is with the persons who submit ghostwritten work as their own for personal gain. This could be dealt with using the laws against fraud, which are already in place. One body has even suggested that going public with details of cases of academic misconduct is itself academic misconduct. The German Rectors' Conference (*Hochschulrektorenkonferenz*, HRK), the voluntary association of state and state-recognized universities in Germany, published recommendations to this nature (Hochschulrektorenkonferenz 2013, II. Empfehlungen, 1.):

> Zum Schutz der Hinweisgeber (Whistle Blower) und der Betroffenen unterliegt die Arbeit der Ombudspersonen höchster Vertraulichkeit. Die Vertraulichkeit ist nicht gegeben, wenn sich der Hinweisgeber mit seinem Verdacht an die Öffentlichkeit wendet. In diesem Fall verstößt er regelmäßig selbst gegen die Regeln der guten wissenschaftlichen Praxis. Dies ist auch bei leichtfertigem Umgang mit Vorwürfen wissenschaftlichen Fehlverhaltens der Fall sowie bei der Erhebung bewusst unrichtiger Vorwürfe (vgl. geplante Ergänzung zu DFG, Sicherung guter wissenschaftlicher Praxis, Empfehlung 17 [...]).

> In order to protect the whistleblower and the person accused of academic misconduct, the work of the ombudsperson must be conducted in utmost confidentiality. The secrecy is not given when the whistleblower makes his case public. In this case, the whistleblower himself is guilty of academic misconduct. This is also the case for frivolous accusations of academic misconduct and purposefully raising untrue accusations (see the planned extension to the

---

[5] The German Association of University Professors and Lecturers (*Deutscher Hochschulverband*, DHV) suggested in 2012 that a new crime be defined, scientific fraud (Deutscher Hochschulverband 2012).

DFG recommendations for securing good academic conduct, number 17).
(translation by the author)

The DFG has not yet published a version in English, but in German they have not followed this exact wording of the HRK (Deutsche Forschungsgemeinschaft 2013). They suggest confidentiality must be maintained if the ombud process is used, and note that it could be construed to be academic misconduct if the press is informed before the university. However, they make no statement on how much time the university is to be given. They also state that the universities do not have to act on anonymous whistleblowers if they do not see a reason for doing so. For a more detailed discussion on this topic, see Weber-Wulff (2013) with links to other online discussions and Heßbrüggen (2013) for a timeline of events and discussions.

Indeed, only focusing on the students as the ones somehow at fault must be seen as only partially addressing the root problem. As a 2013 survey in Germany has shown, while one in five students interviewed over a period of three years admitted that they had plagiarized at least once during the past six months – almost all of those plagiarisms did not have any negative consequences (Sattler, Graeff, & Willen 2013). This seems to be the crux of the matter.

Cheating happens on all levels of academia, and it is not just restricted to Germany. People see this going on, but are afraid to say something for fear of endangering their own careers, as the author has heard countless times in emails telling her about cases of cheating. They want something to be done about it, but are not willing for their own names to be named as the whistleblower, often because they are close to the perpetrator. Perhaps what has happened is that so many people have been seen getting away with cheating and plagiarism that this has lowered the barrier. More and more people do not find it as problematic to justify their own cheating, if "everyone" is doing it.

Now that some people have channeled their irritations with the situation into a public, academic documentation of the plagiarism that exists in published theses, a slightly different perspective appears. VroniPlag Wiki has shown that it is not just the individuals who cheat – the system also makes it easy to cheat, and people who should be in charge of keeping things clean seem to be looking the other way. It is, however, not clear why this is so.

The universities have to educate more and more people with less and less money. They are being held to quantitative measures of quality that at times border on the absurd. Researchers will, of course, do all in their power to maximize whatever factors are being used to dole out money in order to obtain sufficient funding for the research they want to conduct. But apparently, along the way, the quality of teaching and learning has gotten replaced by management data. True quality is hard to quantify, quite particularly in education and research.

Putting more and more students into the courses means there is less time for personal encounters with teachers. The number of people working in *Mittelbau* positions, the young academics working on their doctorates and habilitations in Germany, has risen over the past 10 years, as has the number of students, but a much larger percentage of them are working in research-only, non-permanent jobs (Statistisches Bundesamt 2012). Since much of the funding for research comes from outside sources or takes place in specialized research institutions that do not teach undergraduates, this means that there are fewer people on the level below professor who are actually involved in teaching, So students are more and more left to their own devices for finding, reading, and producing papers, since they are not getting much individual feedback on their work.

Some mentors appear to be more interested in their own private projects than taking time to advise and discuss with students. Few points in internal rankings, if at all, are given for mentoring, only the number of completed theses and passed students counts. And the students do not complain, do not stand up when they see their work being misused by people further up the hierarchy. This helps perpetuate the problem.

It is strange to see that, on the one hand, many academics were agitated about zu Guttenberg's plagiarism and insisted that he step down over this issue, but when it comes to doing something at their own schools, they are unsure how to proceed. The universities and research societies announced action. Or rather, as a cynic would say, they found a few things that could be done that would look like action on the outside, without changing the basic nature of German universities, as mentioned above: They purchased software, insisted that their doctoral students sign a sworn statement that they had listed all their sources, and extolled the virtues of organized doctoral programs.

Such doctoral programs are just the most recent attempt in a centuries-long struggle to get the granting of doctorates under control. As Ulrich Rasche notes (2007, pp. 276–277), in the 17th and 18th century a doctorate was a possibility for commoners to achieve a status similar to that of nobility. In the 17th century the most important part of the dissertation process was not exams or the thesis, but the formal dinner that had to be organized and paid for by the candidate for all of the professors in the faculty in order to proclaim the new doctor and introduce him to the academic community[6].

Rasche states that around the year 1700 universities came up with the notion of a formal doctoral certificate and insisted on levying a high fee for preparing this document, instead of the dinner. The universities could then use the fees as they saw fit. This led, however, to the universities being eager to sell doctorates in order to finance themselves. They even developed a sort of mail-order doctorate production system, the promotion *in absentia*. Not only did the doctoral candidates not have to study at the university in person, but the thesis was often written by the doctoral advisor himself (Rasche 2007, p. 281) and in some cases even the defense was omitted[7].

---

[6] The *prandium doctorale* was so expensive, Rasche describes, that sometimes a number of candidates would pool their resources and hold the dinner together (Rasche 2007, p. 278).

[7] Rasche (2007, p. 291), reports on a case that was quite widely discussed at the time. The University of Greifswald awarded a doctorate in medicine *in absentia* to Johann Peter Menadie in 1774, and did not insist that he defend the thesis. However, Menadie turned out not have been a doctor, although he had stated on his application that he was, but a cobbler, and the thesis had been written by the dean. When Menadie's true profession became known, the university rescinded the doctorate, although it was rather scandalous that he had demonstrated that a doctorate could be purchased.

In the 19th century, new universities such as the University of Berlin (Rasche 2013, pp. 309–310) began to demand that the dissertations be written by the doctoral students themselves and to be published so that they were accessible for everyone and thus subject to discussion. Many universities continued to sell dissertations simply because they needed the money and the candidates wanting to enjoy the particular social status a doctorate confers were willing to pay for them. It was not until 1935 that professors were forbidden from accepting money from their students (Rasche 2013, p. 345).

After World War II, the West German universities were state-funded and the professors were given money to pay the salaries for a few doctoral students. Professors who managed to obtain external funding were able to finance more students. People report, off the record, that in some departments the doctoral students are often busy preparing instruction or doing research for their doctoral advisors and have little time for their own research. They end up patching together a dissertation just before their funding runs out and submit it to the faculty a few weeks in advance of the defense. In theory, all members of the faculty can read and comment on the thesis, as well as attend the defense and ask questions there. In many cases, the theses do not appear to be read very thoroughly, if at all, and thus those candidates who take shortcuts are not discovered prior to the granting of the degree.

Academic misconduct has turned up over and over in many different fields and thus points to a systematic problem in the German academic community at large. These are just a few of the more prominent ones.

- Medical and business researchers have had to retract numerous papers:

  – Joachim Boldt (medicine) has had to file 89 retractions by the end of 2012 (Retraction Watch 2011a).
  – Silvia Bulfone-Paus (medicine) has 13 known retractions (Retraction Watch 2011b).
  – Ulrich Lichtenthaler (business) currently has 12 retractions (Retraction Watch 2013).

- The book produced in honor of the 300-year anniversary of the founding of the Charité hospital in Berlin had to be recalled on account of massive plagiarism (Horstkotte 2010).
- Jan Hendrik Schön, a physicist, was granted a doctorate in 1998 by the University of Konstanz. Schön was fired without notice from Bell Labs because of widespread misconduct and numerous papers having to be withdrawn from top publications, as reported in *Nature* by Brumfiel (2002). The university rescinded Schön's doctorate in 2004 on the basis of his involvement in the repeated publication of fabricated data. He protested against this action, but the university did not accept the protest. Schön took the university to court and won (Verwaltungsgericht Freiburg/Br. 2010). The university then appealed to the upper court (Verwaltungsgerichtshof Baden-Württemberg 9. Senat 2011) and the decision was overturned. Schön appealed that decision to the federal court, but they dismissed the appeal (Bundesverwaltungsgericht 2013).

From the volume of email that the author receives on this topic and the questions asked during public presentations, there seem to be many students and researchers who are aware of plagiarism and academic misconduct problems, but who are unsure of how to proceed with such a case. They are often also worried that if they voice their concerns, they may then themselves face retaliation and could thus endanger their careers.

As noted in Sect. 2.4, the German Research Foundation DFG published guidelines for good academic practice in 1998. Appendix A reproduces the official English translation. Unfortunately, it seems that many students and researchers in Germany are unaware of the existence of these guidelines. Some countries have long had procedures of checks and balances in place for cultivating a culture of good academic practice. Germany – and perhaps other countries – must learn from the experiences of those who have been developing policies and procedures for quite some time. Chapter 6 will give some examples.

As soon as one begins digging deeper into the question of promoting good academic practice, many questions arise. How can good academic practice be taught? What culture must exist in the laboratories so that academic misconduct can be avoided? Is there some specific property of an academic system that provides an environment in which academic misconduct can thrive? What can be done to make it clearer how attribution must be done? This is not just a problem in Germany. Many cases from around the world are showing that plagiarism and academic misconduct are topics that cannot be ignored.

Although much has been written in English on the topic of plagiarism and how to deal with it for students, there is little material about plagiarism in German and almost nothing written about plagiarism in dissertations or by professors in either language. An older work, (Schaltenbrand 1994), discussed quite a number of cases, but does not give much in the way of references in order to find more details. Ingo von Münch, a law professor and a former minister of science and technology in Hamburg, wrote an interesting volume about dissertations in general (Münch 2006) and in 2012, in reaction to the plagiarism cases, a slim volume on good science (Münch 2012). Volker Rieble, a law professor from Munich, published a book in 2010 on plagiarism by professors (Rieble 2010) that had to be quickly taken from the market, as some of the persons named (in particular, other law professors) filed suit against him for defamation of character and won (Horstkotte 2011), although Rieble has filed an appeal that is still pending in 2013. In the area of plagiarism in literature, Philipp Theisohn published an excellent historical overview (Theisohn 2009). On a more humorous note, Roland Schimmel published a 10-step guide for preparing a plagiarized dissertation (2011), with, on a sarcastic note, a foreword purportedly written by zu Guttenberg.

This book cannot go into more detail discussing all that has been published on the topic. But the goal is to provide material for discussing plagiarism on all levels of scholarship – students, doctoral students, researchers, and professors – and to thus encourage universities, not only in Germany, to take action. Before continuing, since readers are not presumed to all know how the German doctoral system works, a brief excursion with some historical context is presented.

3.4 Dissertations in Germany Today

A doctoral dissertation is submitted to a university in Germany today as a partial fulfillment of the requirements for obtaining a doctoral degree. It is expected that the doctoral dissertation contribute some new, previously unpublished material to the research field. The dissertation is usually defended (*Verteidigung*), although some universities require oral exams (*Rigorosum*) either instead of or in addition to the defense. Some universities require coursework to be completed, others will accept a dissertation from anyone who submits a document to the faculty. Each university sets its own rules for conferring a degree.

It is then necessary for the dissertation to be published and placed on record at the German National Library and at a number of university libraries around the country. The publication can either be with a publishing house, or can be self-published, or even done on microfiches or published with an electronic document server. Some universities have even experimented with so-called cumulative dissertations, in which the candidate submits a number of published papers as their dissertation. But with the problems that have arisen around conferences and journals that only exist for the purpose of producing publications, this type of dissertation is increasingly being called into question.

It is extremely desirable for people in Germany to obtain a doctoral degree, as it can be added to one's identification documents and is then used socially, not just in academic circles. Many people enjoy the social reputation that comes with being called *Herr Dr.* or *Frau Dr.*, especially in the area of business and politics. Germans automatically have great respect for people with doctoral degrees. Throughout Europe there are companies that will also pay someone with a doctoral degree more than someone who only has a master's or bachelor's degree, even if they are doing the same job. That makes the degrees highly prized goals.

A medical doctorate is seldom on the same level as doctorates in other fields. Students of medicine finish their studies and are licensed to practice on the basis of a state examination. Formally, if they want to be called Dr., they must also submit a dissertation, although *Doktor* is also the name of the profession. Many theses are completed in parallel with grueling studies and residency requirements, so the theses in general are often thin volumes that may just be a write-up of one experiment or an investigation of a few clinical cases. The shortest one found to date consists of a four-page article with two authors, Khaleghi Ghadiri & Gorji (2004), that was submitted by the first author Khaleghi Ghadiri (2006) to the University of Münster as a medical dissertation two years later. And even though the medical theses are quite short, some doctoral students will reuse text blocks describing an experimental setup or a methodology from other publications of the research group, often without attribution.

The *Wissenschaftsrat*, the German Council of Science and Humanities, is an advisory board for the Federal and State governments concerned with the structure and development of higher education and research. It published some strong words in 2004 about medical dissertations in Germany (Wissenschaftsrat 2004, p. 74–75):

Medizinische Dissertationen und Habilitationen, abgesehen von den auch hier existierenden hervorragenden Arbeiten, erreichen oftmals nicht das wissenschaftliche Niveau, das in anderen Disziplinen üblich ist. [...] Als Ergebnis legte die Studie [gemeint ist (Weihrauch, Starte, & Papst 2003)] damit offen, dass die untersuchten medizinischen Dissertationen von den Kennzahlen her eher Diplomarbeiten in naturwissenschaftlichen Fächern als den dort üblichen Dissertationen entsprachen.

Medical dissertations and habilitations, with the exception of the occasional exceptional work, often do not reach the academic standards of other disciplines [...] as a result the study [(Weihrauch, Starte, & Papst 2003) is meant] demonstrates that medical dissertations are more similar to diploma theses in the natural sciences than the dissertations done in these fields, when one looks at the data [for the amount of effort invested and the publication success].
(translation by the author)

But these words have, unfortunately, not led to any change in the way medical dissertations are prepared.

According to data from the German national statistics office (Statistisches Bundesamt 2012), there were 26,981 doctorates conferred by German universities in 2011. Of this number, 7,771 or almost 29% were in medicine. If one looks at the percentage of medical students graduating in 2011, they made up only 7% of all graduates (24,829 out of 353,839). There were 1,563 completed habilitations in 2011 (a drop of 11% with respect to 2010), with 799 or a whopping 51% in medicine. If the medical dissertations are removed from the number of dissertations and the number of habilitations added back in, the medical theses would be 4% of the total, which would be about commensurate with the proportion of students studying medicine. This suggests that a habilitation in medicine is comparable to a dissertation in other fields. However, there is little real interest in changing the current system within the field of medicine, even though a discussion arises every so often (for example, Spiewak (2011) cites discussions, many going back hundreds of years).

Rescinding a doctorate is a very complicated matter. The body responsible for taking back a doctorate awarded in error is the university that awarded the doctorate in the first place. They will tend to pass the matter on to the school, which will usually convene a committee that is charged with investigating the situation, although the dean can at some universities present the matter directly to the faculty board of the school.

There is one important aspect of German dissertations, as noted above, that assists the detection of plagiarism: They must all be published. This was not always the case in earlier times. The Berlin historian and jurist Theodor Mommsen introduced the notion of compulsory publishing of dissertations in the late 19th century in an article that he published (Mommsen 1876) lambasting the current system that was riddled with corruption and dissertations for sale. The dissertations needed to be available for public scrutiny, Mommsen felt. Ulrich Rasche (2013) gives a thorough overview of the historical situation, Oberbreyer (1878) put together a collection with two articles by Mommsen and much of the contemporary commentary and raging debate from the newspapers and scholarly journals of the time on the topic. The necessity of publishing dissertations has been instrumental in being able to highlight plagiarism, as noted in 1969 by an editor with *German Life and Letters* (Editorial Notices 1969, p. 163):

> It is worth noting, with English conditions in mind, that this particular case only came to light because the dissertation was published and thus came to the notice of a reviewer whose own works had been pillaged. How soon would it have been noticed in this country, where dissertations do not usually appear in print?

This is exactly the point: publishing dissertations makes them widely accessible. Publishing dissertations as open access documents, as is possible today, thus increases the possible readership and therefore increases the chances of finding plagiarism, should it exist.

3.5 Past Plagiarism

The zu Guttenberg doctorate was not at all the first one to be rescinded in Germany. There have apparently been quite a number of degrees in German-speaking areas that were rescinded for plagiarism, and many other cases of academic plagiarism, reaching back for centuries.[8] There are currently over six dozen cases about which at least a smidgeon of information has been collected. This information is public, as it is published either in academic journals or in other periodicals.

Universities, however, especially in recent times, have tended to be extremely discreet in handling the cases. Although much material did not survive the wars, there are an astonishing number of cases for which evidence can be found only by sifting through publicly available material. Some of the more spectacular or well-known cases will be presented in this section.

3.5.1 On Public Notice

The University of Marburg published the notice in Figure 3.2 in 1865 in a German-language academic journal[9].

The dean of the Humanities Faculty, to which the Department of Pharmacy belonged, states that once again they are having to publish the names of people who copied their dissertations. That points to there being other cases in previous years. In this notice two persons are named, W[enzel] Hildwein and A[loys Wilhelm] Josten. The notice gives the titles of the thesis and the names of the authors who were plagiarized. The notice closes with the statement that this should serve as a warning for

---

[8] The author of this book is extremely fortunate that there are a number of expert researchers with the VroniPlag Wiki group who have been combing the digital archives and have uncovered many fascinating cases. A special documentation area is used by the group at HistorioPlag Wiki (2013). The researchers wish to remain anonymous, so I can only thank them profusely without naming them – you know who you are. I could not have done it alone.

[9] This case has been extensively documented in an archived forum at VroniPlag Wiki (2012b).

Bekanntmachung
der
philosophischen Facultät zu Marburg.

Schon wiederholt sind wir in der Lage gewesen, die Namen derjenigen, welche sich durch abgeschriebene Dissertationen auf betrügerische Weise die Doctorwürde zu erschleichen versuchten, der Oeffentlichkeit zu übergeben. Auch in jüngster Zeit sind, trotz unserer bekannten Strenge, wieder zwei derartige Fälle vorgekommen. Es hat nämlich

W. Hildwein, mag. pharm. aus Prag,

seine bei uns eingereichte Dissertation „*De theoriis electrochemicam antecedentibus nonnulla*" beinahe wörtlich aus der gleich betitelten Dissertation des Dr. F. A. Marquidorf, die im Jahre 1843 in Halle erschienen ist, abgeschrieben, und der

Apotheker A. Josten aus Siegburg

die bekannte Schrift F. Müller's „Für Darwin" zu einem wörtlichen oder fast wörtlichen Auszuge für seine Dissertation benutzt.

Zur Strafe für die Betreffenden und zum warnenden Beispiel für Andere werden diese zwei Betrugsversuche hiermit zur öffentlichen Kenntniss gebracht.

Marburg, den 27. März 1865.

Zwenger, z. Decan.

(Zwenger 1865)

Fig. 3.2 Announcement of two plagiarisms in an academic journal from 1865

others and notes that this conduct is to be considered deception, not just some sort of technical mishap.

However, this public notice seems not to have kept either of the gentlemen in question from using the degree. Hildwein later published a number of academic papers. Josten set up a company for producing mineral water and was listed in the Aachen address books for 1877 and 1887 (Adreßbuch 1877, 1887) as Dr. Josten.

3.5.2 André Haas

André Haas (born 1885, date of death unknown) from the Alsatian town of Mulhouse (Mülhausen), passed his oral exams in 1910 and published a thin volume
(Haas 1912) on water management that he submitted to the University of Basel in
Switzerland in 1913 as his dissertation. His advisor was Stephan Bauer. The thesis
consists of 84 pages, but 21 of these are only illustrations, most of which were not
referred to in the text. Subtracting the six pages of appendix and six more pages that
make up the literature list, together with a few empty pages, leaves only around 40
pages of text in the thesis, of which two are half-filled with pictures, one is a quote
in Latin (in the original and translated), and four are quoted from another source.

The plagiarism was apparently discovered not too long after the degree was conferred, as the publication report by the University of Basel (1913) in the next year
already reported that his doctoral degree had been rescinded because it contained
an extensive plagiarism of a published speech (Rehbock 1907) by the hydraulic en-
gineer and professor in Karlsruhe, Theodor Rehbock. Haas had, according to his
CV, at one time been a student of Rehbock's. Curiously, the thesis, although ostensibly about Alsatian water management development, does not actually ever refer to
Alsace, except in his biography, as he is from the area.

It is interesting to examine this plagiarism in the present day. The publication
of Rehbock's speech can be obtained by interlibrary loan and shows that the thesis
indeed includes a copy & paste plagiarism of the speech. For the rest of the thesis,
one can easily enter a few words into Google Books and find additional verbatim
sources for other passages. Section 4.2 will go into more detail on how simple it is
to find plagiarism in this manner.

3.5.3 Max Anton Dietz

Max Anton Dietz (1896–1971) was a dentist in Beilngries in Bavaria who had studied in Würzburg following his military service in World War I. After having worked
as a rural dentist for a number of years, he submitted a slender volume of just 24
pages to the University of Würzburg in 1931 on the topic of Johann Wolfgang von
Goethe's troubles with his teeth (Dietz 1931).

It was the year before the centenary of Goethe's death, and as such there was
much being written about every aspect of Goethe in the press. A theater critic from
Berlin, Hans Knudsen, who was later on to become a professor for dramatics, had
somehow stumbled upon the dissertation and saw that the thesis was a plagiarism
from at least three other literary sources about Goethe.

Knudsen was so angered by this thesis that he felt was both on a silly topic and
bad academic practice to boot that he published a scathing review in a German daily
newspaper (Knudsen 1931). He noted that it would have perhaps been possible to
discuss Goethe's teeth from a dentist's point of view for a proper medical thesis,
but to be taking bits and pieces from non-medical writings, stitching them together

poorly and be given a medical doctorate for his troubles – this was definitely not what a doctorate was supposed to be.

Georg Sticker, the doctoral advisor of Dietz, was a well-known professor in the field of the history of medicine. He joined the fray, publishing a rebuttal in the same newspaper two months later (Sticker 1931). Sticker did not address the plagiarism at all, but instead described why it was perfectly permissible for a dentist to obtain a doctorate on this topic. He also announced that the discussion of dissertations had no place in the daily news, but should be restricted to academic journals.

Quite a number of other articles were written in various scholarly and non-scholarly publications, ranging from daily newspapers over literature journals and library science volumes to dentistry correspondence. Knudsen apparently attempted to get a rebuttal published, but was unable to find a venue. So he put together a 19-page treatise detailing his accusations and including a synopsis of two of the more blatant borrowings (Knudsen 1932). Copies were then sent to many university libraries. One passage from this booklet could just as well have been written today (Knudsen 1932, p. 9):

> „Dissertationen gehören nicht in Tagesblätter". Umlernen, Herr Professor! Die guten Dissertationen kommen längst in Tagesblätter, und die schlechten, wie die von Ihnen angeregte, gebilligte und verteidigte über „Goethes Zahnleiden", gehört erst recht in die Tagespresse, weil das deutsche Publikum ein sehr großes Interesse daran hat, zu erfahren und darüber aufgeklärt zu werden, was für ein Unfug an einer Universität möglich ist und wie es um das Bildungsniveau an der Universität bestellt ist; in einer Zeit vor allem, in der die Regierungen sich genau überlegen müssen, ob und welche Mittel sie für die Aufrechterhaltung von Hochschulen und Lehrstühlen aus Steuer-Mitteln noch erübrigen können oder streichen müssen. Durch diese Art Dissertationen jedenfalls wird keinerlei Berechtigungs-Nachweis geführt. Das große Publikum soll das durchaus erfahren! Die Universitäten haben sich in ihrer Notwendigkeit heute gründlichst zu rechtfertigen.

> "Dissertations do not belong in the daily newspapers!" On the contrary, my dear professor! The good dissertations are already discussed in the daily newspapers, and the bad ones, such as the one you suggested, advised, and accepted about "Goethe's Teeth", must especially be in the papers, because the German public needs to be informed and is interested in the nonsense that is produced at universities, and what level of education is found there; at a time above all in which governments must decide how to allocate taxpayer's money for funding institutions and professorships and what to no longer fund. This kind of a dissertation does not qualify as proof of entitlement to such a privilege. The people at large need to know! The universities must today thoroughly justify the need for their existence.
> *(translation by the author)*

Despite the public support from the doctoral advisor, the doctorate appears to have been rescinded[10] and the case dropped out of sight, there being more pressing issues

---

[10] A notice was published in the *Jahresverzeichnis der an den deutschen Universitäten und Technischen Hochschulen erschienenen Schriften*, Vol. 51 (1935), p. 1124: "Zu Jahresverzeichnis 47. 1931 [...] 7398: Die Dissertation ist eingezogen worden, da sie sich als Plagiat herausgestellt hat." (the dissertation of Max Anton Dietz was retracted on account of being discovered to be a plagiarism). However, most of the records of the University of Würzburg were destroyed just before the end of World War II in a bombing raid that caused a fire in the university archives. They have records of his matriculation and graduation, but no record of the degree being rescinded. It is, however, noted in a number of online catalogues, for example, the *Gemeinsamer Verbundkatalog* at http://gso.gbv.de/DB=2.1/PPNSET?PPN=45103256X.

in Germany in the 1930s and 1940s. Strangely enough, some modern publications about Goethe have for some reason chosen to refer to the thesis submitted by Dietz, without realizing that it was rescinded on account of plagiarism See, for example, (Veil 1946, p. 26), (Theopold 1964, p. 236), and (Ullrich 2002, p. 355), to name only a few. There is even an article in the renowned *Goethe Jahrbuch* (Ullrich 2006, p. 182, Fn. 34) that still refers to the thesis in a footnote, apparently unaware that it is a plagiarism. This is one of the problems that arises with plagiarized dissertations – they enter the body of academic knowledge and are extremely difficult to remove again.

3.5.4 Paul Englisch

The jurist and writer Paul Englisch (1887–1935) was sued in 1929 by the publisher Georg Socher. Socher had purchased a handwritten manuscript from the publisher Eugen Marquardt in 1920 that Marquardt had acquired from the author Bernhard Stern-Szana from Vienna in 1908. Stern-Szana's book was about dirty jokes in literature and arts. Socher had hired Englisch to type up the manuscript in 1922, but never published it. A few years later, Englisch published a book (Englisch 1928a) that Socher felt was a plagiarism of the typescript that Englisch had prepared. Socher prepared extensive material giving the history of the situation and demonstrating the plagiarism for the court case[11].

The documentation shows excerpts from Englisch's book on the left hand side of a large album page and one page of the typescript on the other side, with boxes and arrows pointing to the text portions that are similar. One page is reproduced in Figure 3.3. There appears to have been an appeal in 1932, but the results for neither of the court cases have been found in any archives, they may have been lost in World War II. Englisch subsequently published an identical version of the book with only a new title that same year (Englisch 1928b), and then went on to publish two pamphlets about plagiarism, identifying and documenting the plagiarizing practices of other authors, among them Bertolt Brecht in the Threepenny Opera (Englisch 1930, 1933). He also published a volume about the practice of granting honorary doctorates under his pseudonym Frank Waldassen (1933)[12].

He published a definition of plagiarism in one of his pamphlets (see Sect. 2.1), although he took particular care in insisting that intent must be provable in order for a text parallel to be considered plagiarism. This would, of course, absolve himself from being called a plagiarist in the Socher case. Interestingly enough, there are some plagiarisms in his book about plagiarism that can be found using the digitalizations available on *Google Books* or at the *Visual Library*.

---

[11] Early in 2013 a friend pointed the author of this book to a one-of-a-kind package on sale at a second-hand bookstore: The documentation, a copy of each of Englisch's books from 1928, and a copy of both of Englisch's brochures on plagiarism. The package was purchased and digitized and is now available online (Socher 1929).

[12] Degener (1935) lists Frank Waldassen as a pseudonym of Paul Englisch.

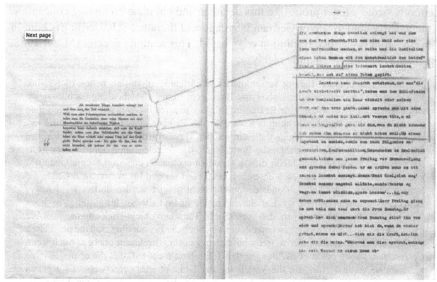

(Socher 1929, pp. 101–102)

Fig. 3.3 Documentation for *Socher v. Englisch*

3.5.5 Father Udo Maria Nix

Father Udo Maria Nix (1927–2000), a Dominican priest, submitted a dissertation entitled *Der mystische Wortschatz Meister Eckharts im Lichte der energetischen Sprachbetrachtung*[13] to the Faculty of Letters of the University of Bonn in 1961. His doctoral advisor, Leo Weisgerber, was a prominent scholar of German. The thesis was graded as very good, *magna cum laude*[14], and the book was published in Weisgerber's series *Sprache und Gemeinschaft* (Nix 1963).

Kurt Ruh, a medievalist scholar of German, reviewed the thesis (1964), complaining eloquently about how substandard it was. A review in the same year by Josef Quint (1964) was even more devastating. The well-known Eckhart scholar Quint found a number of his own phrases in the thesis, and published a precise side-by-side documentation in the review.

As Moser (1968, p. 313) documents, the Faculty of Letters at the University of Bonn decided in 1965 to rescind Nix's doctorate. However, the administrative level above the faculty, the *Concilium decanale*, decided in 1966 not to follow through on that decision, ostensibly for legal reasons. Nineteen professors from the Faculty of

---

[13] The title translates to: "The mystic vocabulary of Master Eckhart in light of the energetic evaluation of language."

[14] According to a letter from Udo Nix to Kurt Ruh, dated 8 February 1965, in which Nix protests against the review. The letter is found in the estate of Kurt Ruh in the Library of the University of Würzburg.

Letters[15] wrote a letter protesting this decision and circulated it among colleagues. At the yearly conference of German language and literature scholars in Bochum in 1967, a strongly-worded resolution was passed. This scathing excerpt (Moser 1968, pp. 313–314) sums it up quite nicely:

> Die auf der Bochumer Tagung versammelten Hochschulgermanisten halten es für ihre Pflicht, sich von dieser an der Universität Bonn getroffenen Entscheidung nachdrücklich zu distanzieren. [...] Wenn eine Rezension in einer unserer Fachzeitschriften gegen eine wissenschaftliche Veröffentlichung den Vorwurf des Plagiats erhebt, hat es als selbstverständlich zu gelten, daß diejenigen, die ein solcher Vorwurf trifft, in angemessener Weise dazu Stellung nehmen. Versuchen die Betroffenen, die Angelegenheit durch bloßes Stillschweigen zu erledigen, und bleibt dieses Verhalten auch noch ungerügt, so muß man fragen, was unser Rezensionswesen eigentlich noch wert sei und bis zu welchem Grade die Regeln wissenschaftlichen Anstands denn außer acht gesetzt werden dürfen. [...] Angesichts einer solchen Häufung von Entlehnungen, wie sie im Falle Nix festzustellen ist, kann weder die Erklärung befriedigen, daß vorsätzliche Täuschung nicht eindeutig nachweisbar und daher bloße Fahrlässigkeit zu unterstellen sei, noch die Behauptung, daß die plagiierten Stellen für die Beurteilung der wissenschaftlichen Leistung irrelevant blieben. Auch wenn sie zuträfen, höben beide Feststellungen den Tatbestand nicht auf, daß die oben genannte wesentliche Voraussetzung für die Verleihung des Doktorgrades irrigerweise als gegeben angenommen wurde. Es wäre schlechthin verderblich, wenn in solchen Fällen die gesetzlichen Vorschriften in einer Weise ausgelegt würden, welche eben diejenigen Grundlagen wissenschaftlicher Forschung und Publikation bedroht, deren Sicherung die gesetzlichen Vorschriften zu dienen haben.

> The scholars of German Letters gathered at the conference in Bochum feel that it is their duty to distance themselves from the decision reached by the University of Bonn. [...] When a review of an academic paper is published in one of our academic periodicals and contains the accusation of plagiarism, it is taken for granted that the person such accused must respond in an appropriate manner. If the person in question tries to solve the matter by remaining silent, and if this behavior is not condemned, then one must ask oneself of what worth our system of reviews actually is and to what degree the rules of good academic conduct may be set aside. [...] In the face of the sheer amount of borrowed material that can be determined in the case of Nix, it is not satisfactory to declare that it is impossible to prove beyond a shadow of doubt that the deception was not done with malice aforethought and thus only an accusation of negligence remains. It is also not satisfactory to assert that the plagiarized passages are irrelevant for the determination of the academic content. Even if this were so – it would not change in the least the fact that the above mentioned preconditions for granting a doctoral degree were erroneously assumed to have existed. It would be ruinous if in such cases the legalities were to be interpreted in such a manner as to threaten the exact same basic tenets of academic research and publication that they purport to uphold.

> *(translation by the author)*

The resolution also notes (Moser 1968, p. 314)[16] that some of the references that Nix did have in his literature list were not correctly given as referring to the original

---

[15] Interestingly, Karl Dietrich Bracher, the doctoral advisor of Margarita Mathiopoulos (see Sect. 3.5.8), was among the signers.

[16] "Abgesehen davon, daß die ohne Kennzeichnung oder doch ohne zureichende Kennzeichnung benutzten (in einem ausgerechnet von P. Nix selbst herausgegebenen Sammelband enthaltenen) Arbeiten von H. Fischer, J. Koch und H. Kunisch im Nixschen Literaturverzeichnis lediglich durch die irreführende Angabe „Nix, Udo und Öchslin, Raphael: Meister Eckhart der Prediger. Festschrift zum Eckhart-Gedenkjahr. Freiburg 1960" vertreten sind, hat das Bonner Concilium decanale damit

sources, but only to texts that were published in a *Festschrift* that Nix had edited together with a fellow Dominican priest in 1960 in celebration of the 700th birthday of Eckhart (Nix & Öchslin 1960).

Both the resolution itself and commentary discussing the situation were printed in the major periodicals of the field, including international journals[17]. But even in the face of all of the protests by the top researchers in the field, the university remained unmoved. As was soon noted (Rosenfeld 1969, p. 3227), if the collective protests of so many professors of German are not sufficient to have a dissertation with over 100 plagiarisms in it withdrawn, then plagiarists have a bright future indeed. Father Nix kept his doctorate and he even authored a number of books for students on effective studying techniques and public speaking while working as an instructor for rhetoric at the Catholic seminary in Paderborn.

3.5.6 Friedrich Wilhelm Prinz von Preußen

According to the German newsweekly *Der Spiegel* (1973), Friedrich Wilhelm Prinz von Preußen (born 1939), the great-grandson of the last German Kaiser and heir to the now defunct Prussian throne, submitted a dissertation in history to the University of Erlangen and was granted a doctorate in 1971.

Spiegel tells the story of a librarian in Marburg who was leafing through the freshly printed volume on Otto von Bismarck, the 19th-century German Chancellor who worked for the unification of many of the independent German states and built up the German Empire (Preußen 1972). Since the librarian was quite interested in Bismarck, he had read many books on the subject. He quickly spotted similarities with other, older works. Upon closer examination, two-thirds of the prince's dissertation was found to be an almost verbatim plagiarism of three sources, with only a few changes that might have been made by a corrector reading the thesis.

The doctoral advisor, Hans-Joachim Schoeps, was informed by the librarian, and Schoeps himself prepared a documentation of the extent of the plagiarism and informed the university, which summarily rescinded the prince's doctorate. *Spiegel* (1973) quotes Schoeps as noting that this was a unparalleled occurrence, as if there had never been plagiarism in a thesis before. Since there were similar utterances about the plagiarism in the thesis of Karl-Theodor zu Guttenberg, it becomes clear that people are quite unaware of all of these previous cases of plagiarism.

In the early 1980s the prince was again conferred a doctorate, this time by the LMU in Munich on the topic of the history of his own family during the 1930s and 1940s (Preußen 1984). He was interviewed about the parallels between the zu

---

eine Entscheidung getroffen, die einen Präzedenzfall schaffen und katastrophale Folgen heraufbeschwören kann."

[17] The resolution is printed in its entirety in Moser (1968); a summary in English is in Editorial Notices (1969, p. 163); Nix's publisher refers to the resolution and notes that they are withdrawing Nix's publication in (Pädagogischer Verlag Schwann 1968).

Guttenberg case and his own experiences (Medick 2011), in particular about how he came to be writing the second thesis.

As an interesting side note, Google Books can be used today to find additional sources for phrases copied word-for-word in the dissertation on Bismarck, for example, Gerhardt & Hubatsch (1950). A list of all phrases taken from this source can be found at the VroniPlag Wiki (2013c).

3.5.7 Elisabeth Ströker

In October 1990, a philosophy professor from the University of Cologne, Marion Soreth (born 1926), published a 411-page book (Soreth 1991) containing a detailed documentation of text parallels in the dissertation of fellow Cologne philosophy professor, Elisabeth Ströker (1928–2000) along with some comments on Ströker's habilitation. The doctoral thesis (Ströker 1953) had been submitted to the University of Bonn in 1953 under the mentorship of Theodor Litt, a cultural and social philosopher. Soreth presented quite a detailed synopsis of extensive fragments from Ströker's dissertation that were not properly marked as being either word-for-word or paraphrased copies from various sources.

Soreth meticulously documented the page and line numbers, and included explanations between the fragments. The book has a number of cross-reference tables, listing the pages of the dissertation and the sources with all of the relevant text parallels collected. It is quite similar to the type of documentation that is prepared in the VroniPlag Wiki in the present day.

A copy of the book was put on the dean's desk, as well as in the mailboxes of Ströker and a few other colleagues, and was available in bookstores around Cologne. The newspapers soon picked up the story (for example, Eisenhut 1990; Meichsner 1990). The reaction of colleagues and of the University of Bonn was, however, rather disturbing. Instead of reading and discussing the documentation in a serious fashion, some people – especially Ströker – attacked the whistleblower. There were speculations about Soreth's motives and there was in general much agitation. In those pre-Internet days, instead of posting on blogs, there were notices put up on bulletin boards around school and anonymous flyers could be found around campus[18].

Ströker tried to obtain an injunction to ban the book, but instead was only able to force Soreth to quickly publish a second edition with eleven lines removed from the epilogue that were considered to be a personal attack on Ströker. Ströker went on sick leave and investigations were started, both in Cologne and in Bonn.

The University of Cologne determined that the plagiarism was substantial, but since the decision to rescind the doctorate could only be taken by the University of Bonn, they deferred to that decision. In Bonn, a committee took the entire semester to deliberate, although the documentation was available in printed form. They did

---

[18] Facsimiles of the flyers, as well as articles from various journals, newspaper clippings, and letters of support can be found in Soreth (1996).

not investigate the plagiarism, however, but instead decided only to focus on legalities (Soreth 1996, p. 47). Even if it was a plagiarism, the doctorate could only be rescinded if deception was involved, the committee determined, although there is no legal definition that says anything about intent. They decided that the referees were not deceived, because naturally they were familiar with the material and would know that this was not Ströker speaking, but the respective authors (Soreth 1996, p. 49).

Of course, anyone else reading this published work would not have that same knowledge basis, and the text parallels were found on almost 50% of the pages. Ströker defended herself by saying that she had submitted a version with more footnotes, but that appeared to have been lost and the footnotes were accidentally removed by the secretaries when preparing the thesis for printing the depository copies. The committee agreed that this could have happened, noting that it was impossible to determine if this was the case so many years later (Soreth 1996, p. 51). And since that was not deception, they decided not to begin proceedings to rescind the doctorate. They did, however, note in their final report[19] that the "quotation culture" used in the thesis did not find the approval of the investigating committee.

Rescinding a doctorate is, of course, an entirely different question than just deciding if a dissertation is a plagiarism or not. It is possible for the university to determine that the academic misconduct would perhaps have been grave enough not to grant the doctorate, if found early enough, but not serious enough to take back the degree. This does beg the question as to how the academic community is to treat a thesis that contains plagiarism, but has not been withdrawn.

Ströker continued to teach at the University of Cologne until she retired, and was often to be found lamenting in some form or other that Soreth had destroyed her (Ströker's) career through her "mud-slinging." Just before her death, she published a long-winded book (Ströker 2000) complaining about all the problems that had befallen her. However, it was the reputation of the whistleblower, Soreth, that was tarnished.

Over 100 persons, professors of philosophy and publishers, signed an open letter (Acham et al. 1991) that was published in a professional journal denouncing Soreth for the manner in which she had published the allegations about Ströker. They noted that they had worked together with Ströker over the past years and found her to be a fine academic. According to Soreth (1996, p. 99), some of the signers did not have a copy of the text of the open letter available, but just gave their name over the telephone, and some did not even read the documentation, they were just upset by the press reports they had read.

Soreth did find a few supporters, and she published material from them together with anonymous hate mail and bulletin board notices such as in Figure 3.4, as well as the newspaper articles that appeared about the case (Soreth 1996). She was, by her own accounts, more or less shunned by her peers, although she continued teaching far past retirement age.

---

[19] The complete final report is documented in Soreth (1996, pp. 46–55) and available online at Kommission Bonn (1991b).

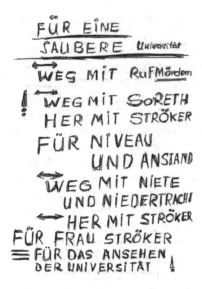

(Soreth 1996, pp. 12–13)

Fig. 3.4 Two of the anonymous hate letters posted on the bulletin board at the University of Cologne (digitally cleaned)

Finetti & Himmelrath (1999, p. 100) quote Peter Weingart, a professor for Science and the Media at the University of Bielefeld, whom they interviewed in 1996 on the topic, as saying that this is an exemplary case of the conservatism of scholars in Germany and of their reluctance to take action against deception and fabrication in their own ranks.

This case perhaps demonstrates that the myth of the self-cleaning powers of academic endeavor is just that: a myth. The German academic community has a tendency to find fault with the person pointing out problems of good academic practice, instead of the persons involved in the bad practice, if those so accused are in a position of power.

3.5.8 Margarita Mathiopoulos

Margarita Mathiopoulos (born 1956) submitted a dissertation to the University of Bonn in 1986 that was published in German (Mathiopoulos 1987) and as an English translation (1989) by renowned publishers[20]. Her doctoral advisor, Karl Dietrich Bracher, a well-known German political scientist who wrote extensively on the

---

[20] An extensive documentation of this case can be found online (MMDoku Wiki n.d.).

Weimar Republic and Nazi Germany, supplied the foreword for the German publication; Gordon A. Craig, a renowned Scottish-American historian also specializing on that period of German history, wrote the foreword to the English translation.

Mathiopoulos, born in Germany of Greek immigrant parents, was nominated by former chancellor Willy Brandt in early 1987, shortly after defending her dissertation, to be the speaker of the German Social Democratic Party. A bitter fight broke out within the party, mostly because she was not a member of the party and had not worked her way up from the rank and file. She withdrew her application, but Brandt eventually was forced to step down as party leader because of all the bickering (Lorenz 2012, p. 217).

Andreas Falke, who was working at the American Embassy in Bonn at the end of the 1980s and is now a professor for American Studies in Germany, was on leave and doing research at Harvard University. He was asked to review her thesis for the Harvard journal *German Politics and Society* (Falke 2012). Falke picked up on the plagiarism and the uneven tone found in the book, wavering "between pseudo-scholarly rhetoric and the style of politicians' Sunday speeches [...]" (Falke 1989, p. 100). The journal delayed publishing the review, as there were concerns about upsetting important people. It was finally decided that the review would be published, but the German newsweekly *Der Spiegel* came out with an article about the plagiarism (*Der Spiegel* 1989) just before the review appeared (Falke 2012, p. 468).

A scandal erupted, accusing Mathiopoulos of plagiarism. Additional articles and letters to the editor were printed in *Der Spiegel* and other media, while more scholarly reviews were also forthcoming. Under pressure, the University of Bonn set up another committee, parallel to the Ströker case, in October 1990 to investigate the allegations. The committee submitted their report in 1991, stating that even though there were technical weaknesses to be found in the thesis, they recommended that the doctorate not be revoked because there was still a "substantial core" (Kommission Bonn 1991a) to be found in the thesis. The faculty board followed this recommendation. Another review (Shell 1991) that was published shortly thereafter noted, however, that no matter if there was plagiarism involved or not, the thesis was so weak that it should not have been accepted.

Mathiopoulos has since been awarded two honorary professorships, one in 1995 from the Technical University of Braunschweig and one in 2002 from the University of Potsdam. At least one of the referees, Gert Krell, had advised the TU Braunschweig not to grant her the honorary professorship on the basis of scholarly weaknesses in her work. He has since made his reasoning public (Krell 2011).

In 2011, the case was reinvestigated by some researchers who documented their results at the VroniPlag Wiki (2011a). First, the seven sources for the text parallels on 33 pages identified by the previous reviewers were closely examined. The researchers now found 62 fragments from these sources, and then identified over 200 additional fragments from twelve new sources, including the dissertation of Friedbert Pflüger, a German politician with the CDU who also completed his dissertation with Prof. Bracher and who was married to Mathiopoulos from 1987 to 2006. Bizarrely, among the new fragments were some that were taken from works of both Bracher and Craig.

The University of Bonn reopened the investigation in July 2011 and rescinded the doctorate on the basis of the new findings in April 2012. In December 2012 a court rejected Mathiopoulos' legal appeal (Verwaltungsgericht Köln 2012b), referring to extensive prior case law on the question of plagiarism and noting that since new sources had been found, it was permissible for the university to again examine the thesis. The courts in Germany have been quite clear on this, upholding the universities' stand on such questions. The court explicitly ruled out the possibility of appeal, but she has filed suit against not being able to appeal to an upper court in Münster (Köhl 2013). That court has still not decided how to proceed; until the decree is final, she is still able to use the doctorate. Should the doctorate remain rescinded, the Universities of Potsdam and Braunschweig have announced that they will also withdraw the honorary professorships that have been conferred on her.

A discussion of the entire case, including some interesting insight into details, can be found in an article in *Amerikastudien* (Falke 2012). Apparently, attempts were made to get Andreas Falke fired from his job at the American embassy. He was also pressured to avoid the use of the word "plagiarism", but instead to speak of "heavy borrowing" in his published review.

3.5.9 Hans-Werner Gottinger

The case of Hans-Werner Gottinger (born 1943) is a very disturbing one, because it demonstrates that the German system of putting the responsibility for dealing with academic misconduct under the auspices of the universities leaves a wide-open problem. How are cases to be dealt with that involve people who are not employees of any university or formal research organization, but are active, publishing academics and are found to have plagiarized? Some researchers give their own company as an institutional affiliation, using a name for the company that suggests a university institution.

Hans-Werner Gottinger is what can be called a serial plagiarist. He has had nine papers retracted to date on grounds of plagiarism. There is a table listing the retractions at (GottiPlag Wiki 2011a); four more papers were discovered to be plagiarisms before they were published (GottiPlag Wiki 2011b). He has claimed affiliation with numerous institutions (GottiPlag Wiki 2011c), but when journal editors have attempted to contact an institution about a retraction, they often discover that he does not actually work there or that he had only worked there sometime in the past. So there is no one to investigate or sanction a case of academic misconduct.

Frustration sets in when there is no way of putting pressure on a person to quit submitting plagiarisms. Editors of two high-profile journals, *Nature* and *Research Policy*, have published articles about the problem and have asked pointedly how Germany plans to get this kind of behavior to stop (Abbott et al. 2007; Martin et al. 2007; Abbott 2008), as Gottinger continues publishing books and submitting articles to journals. In fact, for sometime he was listed as the editor of the open access journal *Open Business Journal* and published at least two plagiarized articles there. For more details, see (GottiPlag Wiki 2011d).

Since there was no reaction coming from the German research establishment, the author of this book wrote to the Minister for Education and Research in Germany at that time, Annette Schavan, at the request of the editors. At first there was no response. After some prodding by telephone an answer was sent: The federal minister has no powers, as the states are responsible for academic matters, but she trusted in the self-cleansing powers of the academic world.

3.5.10 Annette Schavan

Annette Schavan (born 1955) herself, it turns out, also had a plagiarism problem. The German federal minister for Education and Research had submitted a dissertation back in 1980 to the University of Düsseldorf (Schavan 1980). The title translates to "Character and conscience: Studies on the conditions, necessities, and demands on the development of conscience in the present day." At that time, it was possible to directly submit a doctoral dissertation in some fields without first obtaining a lower degree, and she did not have any lower degree.

In 2012 a blogger using the pseudonym of Robert Schmidt published a documentation of 97 possible plagiarisms from 45 sources in the thesis (Schmidt 2012). Twenty-one of the fragments documented are from the habilitation of the Polish Franciscan monk Antoni Jozafat Nowak, according to the blog. The documentation had begun on the VroniPlag Wiki platform in December 2011, but after a few months' work the group decided by majority vote that it was not possible to easily demonstrate the plagiarism in an understandable manner, so the name was not made public, although the documentation remains online (VroniPlag Wiki 2012a).

There was much discussion during 2012 in Germany as to whether the contributors to the VroniPlag Wiki were in some way going easy on Schavan because she was a minister, although it was rather the other way around. If she had not been a minister, the name would not have been made public with just this number of fragments, some of which were quite small. Most of the fragments were not word-for-word plagiarisms and thus were difficult to document so that the general public could understand, although many fragments were obvious to someone familiar with the material. Thus, Schavan was treated exactly the same as a non-well-known person would have been: not named with this amount and kind of documentation. Robert Schmidt continued the documentation and then subsequently found many more fragments.

The University of Düsseldorf conducted a long investigation. During this time quite an extensive debate took place in the media, especially as an expertise was leaked to the press in October 2012. One of the topics of debate was the question of the responsibilities of the advisors for identifying plagiarisms. Another topic was whether there was a statute of limitations on plagiarism, as if somehow the plagiarism could be "healed" by the passage of time alone. Once the expertise was made public, there was a call for an additional, external expertise to be prepared, because it was exceedingly clear from the leaked one that there was no other option than to call the thesis a plagiarism, although the question of possible sanctions is indeed separate from the question of plagiarism.

Many professors and organizations (interestingly, many of them are dependent on financing from the Ministry of Education and Research) denounced the University of Düsseldorf for various supposed infractions, although the university had a legal expertise (Gärditz 2013) demonstrating that the correct procedure was being followed[21]. In February 2013, the university decided that the thesis was indeed plagiarized and then voted as a sanction to rescind her doctorate. As she had no other university degree, her highest education level was now only a high school diploma. She stepped down as German education minister a few days later, stating as her official reason that she could not be education minister at the same time that she was personally taking a university to court.

3.6 Magnitude of Dissertation Plagiarism

Section 3.5 has given just a few examples of dissertations that were rescinded at universities throughout Germany in the past, in addition to the cases that the VroniPlag Wiki has documented. The question often arises as to the magnitude of the problem. It might be expected that with the rise of the Internet the number of plagiarized works might increase. The problem is that there are no numbers available on the number of plagiarisms found, much less on the prevalence of plagiarism, which in principle cannot be measured. It is never clear if people are being truthful when they self-report not having plagiarized.

Although statistics on the number of doctorates granted in Germany are recorded on a yearly basis and are available online, broken down by all sorts of criteria (Statistisches Bundesamt 2011), no central registry of the number of doctorates rescinded exists, although some have been noted in the corrections section of the yearly reports on dissertations completed in Germany (Jahresverzeichnis n.d.) that were published between 1887 and 1987, until the sheer volume of dissertations became unmanageable. Today this information is rather sporadically kept in online union catalogs. There is also some discussion about how this should be done. Some libraries will just remove the note that the book in question was a thesis, others will put information about the revocation in a comment field or annotate the entry about the book being a thesis. Others, like one of the state university library networks, prefer to record precisely when the doctorate was granted and when it was rescinded and have

---

[21] The University of Düsseldorf published a web page with links to two of the more outspoken public criticisms of the handling of the case along with open letters from the rector, Michael Piper, in response to each (Piper 2013). There is a good description of the process the university followed in the press release the university published after the decision (Kohl 2013). An article in the Berlin daily newspaper *Der Tagesspiegel* by Gerhard Dannemann (2013) sums up the entire case quite concisely.

exact rules on how to do this (Hochschulbibliothekszentrum des Landes Nordrhein-Westfalen 2013, Z § 162,9 *Hochschulschriftenvermerk bei Plagiaten*). But since there is no consistent way of noting this across library information systems, it is impossible to automatically gather statistics.

Clues to rescinded doctorates today can mostly be found in various places such as answers from parliamentary questioning by the political parties or buried in various reports of universities or ombudspersons. For example, the state of Baden-Württemberg reported on a number of doctorates that were rescinded in answer to a parliamentary question (Landtag Baden-Württemberg 2010), in 2007 the *Frankfurter Allgemeine Zeitung* reported on one case (Balzter 2007), and *Main-Netz* reports that four doctorates were rescinded at the University of Marburg between 2004 and 2013 (Trauner 2013).

Searching through the legal databases one can turn up occasional cases in which people who have lost their doctorates have sued the universities. In almost all of the cases, the universities have won. Occasionally, the universities have lost the suit because of procedural violations, so they only had to repeat the process, making sure to not make errors the second time.

The cases that have come to light are just the tip of the iceberg. And there is the additional problem that occurs when a dissertation that is later revoked has been quoted by other researchers. Attempting to trace the reference may then land in a dead end, as the reference is no longer available in some libraries, and in others it is not clear that it is a plagiarism at all. It is vital for libraries to keep the books and to mark both the book and the bibliographic entry as being plagiarisms.

3.7 The Problem of Plagiarized Dissertations

On paper, of course, there is no problem. The German research council DFG has its rules about good academic practice, and the universities have rules about dissertations and how they are to be granted. But the rules seem to be ignored occasionally, to the detriment of science. This is distressing for people who see what is going on. Some then bend the rules themselves, and thus are not interested in the rules being followed. Some become disillusioned and leave academia, settling on a career in industry. Sometimes, more care seems to be taken to protect the accused than the persons pointing out the problems.

The problem is that there is no nationwide, transparent process. Each university has different rules for how to set up a committee to investigate the accusations and what body must finally decide on the case. Since this work is in addition to the regular duties of a university professor, unless the media is bearing down on the university, the committee may spend long months deliberating, perhaps hoping that grass will grow over the case or that they can find a good reason to avoid rescinding a doctorate. They might determine that the doctoral candidate did not really mean to plagiarize or they find some legal loophole to keep from having to deal with the problem, perhaps because they are afraid of legal action being threatened.

But even if the cases were dealt with promptly and discreetly, a larger problem remains: each case demonstrates that the procedures for determining good academic practice before granting a degree have failed. Any plagiarism in a published dissertation is one too many. Doctoral students who do not understand what constitutes plagiarism are a danger to the future of science if they continue to work in this plagiarizing way.

There are further questions that arise in connection with rescinded doctorates. Should the names of the offenders be made public? The doctorate is awarded publicly, so one that has been rescinded should be announced so that the academic community is aware that the thesis is no longer valid. In one of the VroniPlag Wiki cases, the thesis turned out to have been previously rejected by another university on the grounds of plagiarism. Would the granting institution have looked more closely at the thesis if it had been public knowledge that a thesis on the same topic by the same person had previously been rejected? How is a university to know that the degree has been previously granted and rescinded or the thesis rejected?

Looking back over the entire process of obtaining a doctorate in Germany, a certain lack of control is clearly observable. Not all universities require their doctoral students to register when they start a doctorate, so there are no national numbers on how many people even start writing a doctoral thesis, only on how many actually complete one. Each faculty, sometimes each professor, determines what the student has to do and how extensive the thesis needs to be, although some universities are requiring agreements to be signed between student and advisor outlining the responsibilities of each party.

There is usually no need for the doctoral student to attend ethics or writing seminars. The mentoring professor is also the grader, together with a colleague, and the other colleagues have so much work of their own that they often do not have a hard look at the output of their peers. And when a problematic case arises, there is often no transparent process to be followed, despite a codified *Promotionsordnung*, a policy each university will have on how it organizes the process of granting doctoral degrees.

How can the quality of this process be improved? How can it move beyond the whims of individual professors to be clear and verifiable? How can whistleblowers be protected? How can the concept of good academic practice be taught better? These are questions that are in urgent need of discussion and action, in Germany and in other countries. The problems run deep. It is important for educators on all levels to understand what compromises plagiarism and academic misconduct and to take action.

References

Abbott, A. with additional reporting by Cyranoski, D., Feresin, E., & Lenotti, C. (2007). Academic accused of living on borrowed lines. *Nature*, Vol. 448, 9 August, pp. 632–633.

Abbott, A. (2008). The Fraudster Returns. *Nature*, Vol. 452, 10 April, p. 672.

Acham, K. and over 100 signatories. (1991). Offener Brief an den Herausgeber der "Information Philosophie", Herrn Peter Moser. In: *Information Philosophie*, Vol. 19, Nr. 3, pp. 40–42.

Adreßbuch für Aachen und Burtscheid 1877/78. (1877). Vogelgesang, E. (Ed.). Aachen. Database with transcribed addresses available at http://wiki-de.genealogy.net/Adressbuch_278 cited 29 July 2013.

Adreßbuch für Aachen und Burtscheid 1887. (1887). Aachen: J. Stercken. Digitized version of a reprint from 1998 available at http://wiki-de.genealogy.net/Adressbuch_228 cited 29 July 2013.

ARD. (2013). *Tatort: Hinkebein*. [Television show], first broadcast 11 March 2012. http://www.daserste.de/unterhaltung/krimi/tatort/sendung/2012/hinkebein-100.html cited 12 July 2013.

Balzter, S. (2007). Aus der Praxis der Plagiatoren, *Frankfurter Allgemeine Zeitung*, 23 October. Available at http://www.faz.net/aktuell/beruf-chance/campus/geistiges-eigentum-aus-der-praxis-der-plagiatoren-1490761.html cited 30 May 2013.

Bewarder, M. (2011). Guttenbergs Plagiate waren schon früher bekannt. In: *Berliner Morgenpost*. 27 November. Available at http://www.morgenpost.de/politik/article1838156/Guttenbergs-Plagiate-waren-schon-frueher-bekannt.html cited 30 June 2013.

Bossemeyer, S. (2011). *Vorwürfe von VroniPlag gegen Dr. Sensburg sind nicht haltbar*. [Press release], 22 December. Available at http://www.fernuni-hagen.de/universitaet/aktuelles/2011/12/22-am-sensburg.shtml cited 17 July 2013.

Brandenburgische Technische Universität Cottbus. (2003). Satzung zur Sicherung guter wissenschaftlicher Praxis an der Brandenburgischen Technischen Universität Cottbus (WissPraxSa) vom 05. Februar 2003. In: *Mitteilungen: Amtsblatt der BTU Cottbus*, 02/2003, 9 March. [University regulation]. http://opus.kobv.de/btu/volltexte/2007/246/pdf/abl02_03.pdf cited 30 June 2013.

Brandenburgische Technische Universität Cottbus. (2012). *Kein Plagiat in der Doktorarbeit von Prof. Dähnert*. [Press release], 12 June. https://www.tu-cottbus.de/btu/de/universitaet/presse/presseinformationen/archiv/details/kein-plagiat-in-der-doktorarbeit-von-prof-daehnert.html cited 21 July 2013.

Brumfiel, G. (2002). Physicist found guilty of misconduct: Bell Labs dismisses young nanotechnologist for falsifying data. In: *Nature: News*, 26 September. Available at http://www.nature.com/news/2002/020926/full/news020923-9.html cited 30 July 2013.

Bundesverwaltungsgericht. (2013). *BVerwG 6 C 9.12, Entziehung des redlich erworbenen Doktorgrades bei späterer Unwürdigkeit wegen Manipulation und Fälschung von Forschungsergebnissen rechtmäßig*. [Press release]. No. 56, 31 July. Available at http://www.bverwg.de/presse/pressemitteilungen/pressemitteilung.php?jahr=2013&nr=56 cited 1 August 2013.

Dannemann, G. (2013). Die Ex-Ministerin und ihre Unterstützer: Schavanzentrisches Weltbild. In: *Der Tagesspiegel*, 3 March. Available at http://www.tagesspiegel.de/wissen/die-ex-ministerin-und-ihre-unterstuetzer-schavanzentrisches-weltbild/7863836.html cited 2 July 2013.

Degener, H. A. L. (Ed.) (1935). *Wer ist's? Unsere Zeitgenossen.* (10th ed.) Berlin: Herrmann Degener.

Deutsche Forschungsgemeinschaft. (2013). *Jahrespressekonferenz 2013: Elektronische Pressemappe zur Jahrespressekonferenz der DFG in Berlin.* [Online press packet]. Available at http://www.dfg.de/dfg_profil/reden_stellungnahmen/2013/130704_jahrespressekonferenz/ cited 5 July 2013.

Deutscher Hochschulverband. (2012). *DHV für Einführung eines Straftatbestandes Wissenschaftsbetrug.* [Press release], 6 August. Available at http://www.hochschulverband.de/cms1/pressemitteilung+M56c11ee1774.html cited 2 July 2013.

Dietz, M. A. (1931). *Goethes Zahnleiden und Zahnärzte.* Volkach a. M.: Karl Hart.

Editorial Notices. (1969). Fr Udo Nix [Editorial]. In: *German Life and Letters,* Vol. 22, No. 2, p. 163.

Eisenhut, L.-P. (1990). Prädikat: „Völlig wertlos". Eklat an der Universität: Professorin wirft Professorin Plagiat vor. In: *Kölner Stadt-Anzeiger,* No. 243/13, 17 Oktober.

Englisch, P. (1928a). *Das skatologische Element in Literatur, Kunst und Volksleben.* Stuttgart: Püttmann. Digitized version available at http://visuallibrary.net/ihd/content/titleinfo/121354 cited 22 August 2013.

Englisch, P. (1928b). *Anrüchiges und Allzumenschliches – Einblicke in das Kapitel Pfui.* Stuttgart: Püttmann. Digitized version available at http://visuallibrary.net/s2wpihd4/content/titleinfo/121905 cited 22 August 2013.

Englisch, P. (1930). *Plagiat!! Plagiat!!: eine Rundschau.* Berlin: Roll. Digitized version available at http://visuallibrary.net/ihd/content/titleinfo/116481 cited 22 August 2013.

Englisch, P. (1933). *Meister des Plagiats – oder die Kunst der Abschriftstellerei.* Hannibal-Verlag: Berlin-Karlshorst. Digitized version available at http://visuallibrary.net/ihd/content/titleinfo/122128 cited 22 August 2013.

Falke, A. (1989). Review. In: *German Politics and Society,* Issue 18, pp. 93–101. Documented in excerpts at http://de.mmdoku.wikia.com/wiki/MMDoku/Dokumentation#Rezension_Falke_1989 cited 19 January 2013.

Falke, A. (2012). Der Fall Mathiopoulos. In: *Amerikastudien.* Vol. 57, No. 3, pp. 467–471.

Finetti, M. & Himmelrath, A. (1999). *Der Sündenfall: Betrug und Fälschung in der deutschen Wissenschaft.* Stuttgart: J. Raabe Verlag.

Fischer-Lescano, A. (2011). Karl-Theodor Frhr. zu Guttenberg, Verfassung und Verfassungsvertrag. Konstitutionelle Entwicklungsstufen in den USA und der EU. In: *Kritische Justiz,* No. 1, pp. 112–119. Available at http://www.kj.nomos.de/fileadmin/kj/doc/zu_guttenberg.pdf cited 10 March 2013.

Fittkau, L. (2013). *Begeisterung für selbst gewählte Worte wecken: Tagung an der Uni Mainz will studentisches Plagiieren bekämpfen.* [Radio interview]. Available at http://www.dradio.de/dlf/sendungen/campus/2181181/. cited 22 July 2013.

Gärditz, K. F. (2013). Gutachtliche Stellungnahme zum Verfahrensablauf in der Sache betreffend Professor Dr. Annette Schavan. Available at http://www.uni-duesseldorf.de/home/fileadmin/redaktion/Oeffentliche_Medien/Presse/Pressemeldungen/Dokumente/Gutachten_Gaerditz.pdf cited 2 July 2013.

Gerhardt, M. & Hubatsch, W. (1950). *Deutschland und Skandinavien im Wandel der Jahrhunderte*. Bonn: Röhrscheid.

GottiPlag Wiki. (2011a). Publications by HW Gottinger involving plagiarism. [Web page] http://de.gottiplag.wikia.com/wiki/Publications_by_HW_Gottinger_involving_plagiarism cited 29 July 2013.

GottiPlag Wiki. (2011b). Papers of HW Gottinger that were found to involve plagiarism *before* being published. [Web page]. http://de.gottiplag.wikia.com/wiki/Plagiarism_before_being_published cited 29 July 2013.

GottiPlag Wiki. (2011c). Institutions with which Professor Hans Werner Gottinger has claimed an affiliation. [Web page]. http://de.gottiplag.wikia.com/wiki/Institutional_affiliations cited 29 July 2013.

GottiPlag Wiki. (2011d). New publications by HW Gottinger. [Web page]. http://de.gottiplag.wikia.com/wiki/New_publications_by_HW_Gottinger cited 29 July 2013.

Guttenberg, K.-T., Freiherr von und zu. (2009) *Verfassung und Verfassungsvertrag: Konstitutionelle Entwicklungsstufen in den USA und der EU*. Berlin: Duncker & Humblot.

Guttenberg, K.-T. Freiherr von und zu, interviewed by di Lorenzo, G. (2011). *Vorerst gescheitert: Wie Karl-Theodor zu Guttenberg seine Zukunft sieht*. Freiburg i. Br.: Herder.

GuttenPlag Wiki. (2011a). *Eine kritische Auseinandersetzung mit der Dissertation von Karl-Theodor Freiherr zu Guttenberg: Verfassung und Verfassungsvertrag. Konstitutionelle Entwicklungsstufen in den USA und der EU*. [Web site]. http://de.guttenplag.wikia.com/wiki/GuttenPlag_Wiki cited 22 August 2013.

GuttenPlag Wiki. (2011b). *Guttenberg-2006/Quellen*. [Web page]. http://de.guttenplag.wikia.com/wiki/Guttenberg-2006/Quellen cited 22 August 2013.

Haas, A. (1912). *Entwickelungsfragen der Wasserwirtschaft in Frankreich und Deutschland insbesondere im Elsass*. Mulhouse (Mülhausen i. E.): Lithographie Marc Geismar.

Häberle, P. (2011). Unvorstellbare Mängel. In: *Süddeutsche Zeitung*, 1 March, p. 13. Available at http://www.sueddeutsche.de/politik/peter-haeberle-zu-plagiats-affaere-doktorvater-revidiert-urteil-ueber-guttenberg-1.1066108 cited 15 August 2013.

Heßbrüggen, S. (2013). Wissenschaftsfreiheit? Deine Mudder … In: *Carta*. [Blog], 14 July. http://www.carta.info/60937/wissenschaftsfreiheit-deine-mudder/ cited 15 July 2013.

HistorioPlag Wiki. (2013). *Hauptseite*. [Web site]. http://de.historioplag.wikia.com/wiki/HistorioPlag_Wiki cited 22 August 2013.

Hochschulbibliothekszentrum des Landes Nordrhein-Westfalen. (2013). *Verbundauslegung der RAK-WB und der RAK-Sonderregeln*. 1 February.

http://www.hbz-nrw.de/dokumentencenter/produkte/verbunddatenbank/aktuell/
verbundvereinbarungen/VK_komplett.pdf cited 3 August 2013.

Hochschulrektorenkonferenz. (2013). *Gute wissenschaftliche Praxis an deutschen Hochschulen: Empfehlung der 14. Mitgliederversammlung der HRK am 14. Mai 2013 in Nürnberg*. http://www.hrk.de/uploads/tx_szconvention/Empfehlung_GutewissenschaftlichePraxis_14052013_02.pdf cited 2 July 2013.

Horstkotte, H. (2010). Berliner Plagiat-Panne: Im Frisiersalon der Wissenschaft. In: *Spiegel-Online*, 2 December. http://www.spiegel.de/unispiegel/studium/berliner-plagiat-panne-im-frisiersalon-der-wissenschaft-a-731186.html cited 30 June 2013.

Horstkotte, H. (2011). Im Streit um Meinungen oder Tatsachen. In: *Legal Tribune Online*, 25 January. http://www.lto.de/recht/hintergruende/h/wissenschaftsplagiate-im-streit-um-meinungen-oder-tatsachen/ cited 30 July 2013.

Jahresverzeichnis. (n.d.). The series has changed names often:
1887–1913: *Jahresverzeichnis der an den deutschen Universitäten erschienenen Schriften*;
1914–1925: *Jahresverzeichnis der an den deutschen Universitäten und Technischen Hochschulen erschienenen Schriften*;
1926–1936: *Jahres-Verzeichnis der an den deutschen Universitäten und Hochschulen erschienenen Schriften*;
1936–1978: *Jahresverzeichnis der deutschen Hochschulschriften*;
1978–1987: *Jahresverzeichnis der Hochschulschriften der DDR, der BRD und Westberlins*

Khaleghi Ghadiri, M. (2006). Natural remedies for impotence in medieval Persia. [Submitted as a doctoral dissertation to the University of Münster (Westfalen) in 2006. Consists of one journal article, pp. 80–83, and a summary in German.] [Entry in the German National Library: http://d-nb.info/982547722, at the University Library in Münster: http://superfix.uni-muenster.de/Katalog/start.do?Query=0="3647285"]

Khaleghi Ghadiri, M. & Gorji, A. (2004). Natural remedies for impotence in medieval Persia. In: *International Journal of Impotence Research*, Vol. 16, pp. 80–83. Available at http://www.nature.com/ijir/journal/v16/n1/full/3901153a.html cited 30 July 2013.

Knudsen, H. (1931). „Goethes Zahnleiden" – Auch ein Beitrag zum Goethe-Jahr und zur heutigen Bildung. In: *Deutsche Allgemeine Zeitung*, Vol. 70, No. 382, 21 August, Evening edition, p. 1.

Knudsen, H. (1932). *„Goethes Zahnleiden und Zahnärzte", die deutsche Bildung und der Würzburger zahnärztliche Doktor-Titel*. Berlin, printed as manuscript.

Kohl, J. (2013). Hintergrundinformationen zum Verfahren zur Überprüfung der Promotion von Prof. Dr. Dr. h.c. mult. Annette Schavan. [Press release], 12 February. Available at http://www.uni-duesseldorf.de/home/startseite/news-detailansicht/article/hintergrundinformationen-zum-verfahren-zur-ueberpruefung-der-promotion-von-prof-dr-dr-hc-mult.html cited 2 July 2013.

Köhl, B. (2013). Mathiopoulos – Noch keine Entscheidung im Streit um Doktortitel. In: *General-Anzeiger*, 31 July. Available at http://www.general-
anzeiger-bonn.de/bonn/Mathiopoulos-Noch-keine-Entscheidung-im-Streit-um-
Doktortitel-article1110283.html cited 1 August 2013.

Kommission in Sachen Doktorgradentziehung der Universität Bonn. (1991a).
*Bericht der Kommission in Sachen Doktorgradentziehung Dr. Margarita
Mathiopoulos.* [Documented at http://de.mmdoku.wikia.com/wiki/MMDoku/
Dokumentation/1991/Bericht_Mathiopoulos-Kommission]   cited 19 January
2013.

Kommission in Sachen Doktorgradentziehung der Universität Bonn. (1991b). *Bericht der Kommission in Sachen Doktorgradentziehung Professor Dr. Ströker.*
[Documented at http://de.mmdoku.wikia.com/wiki/MMDoku/Dokumentation/
1991/Bericht_Str%C3%B6ker-Kommission] cited 22 August 2013.

Kommission „Selbstkontrolle in der Wissenschaft" der Universität Bayreuth. (2011) *Bericht an die Hochschulleitung der Universität Bayreuth
aus Anlass der Untersuchung des Verdachts wissenschaftlichen Fehlverhaltens von Herrn Karl-Theodor Freiherr zu Guttenberg.* Bayreuth. 5 May.
Available at http://www.uni-bayreuth.de/presse/Aktuelle-Infos/2011/Bericht_
der_Kommission_m__Anlagen_10_5_2011_.pdf cited 21 July 2013.

Krell, G. (2011). Wieder ein böses Ende für eine Dissertation? http://www.gert-
krell.de/langfassungmathiopoulos.pdf cited 19 January 2013.

Landtag von Baden-Württemberg 14. Wahlperiode. (2010). *Antwort auf Kleine An-
frage, Drucksache 14 / 7203,* 19 November. Available online at http://www.
theresia-bauer.de/downloads/AA_14_7203.pdf cited 2 July 2013.

Lorenz, E. (2012). *Willy Brandt: Deutscher – Europäer – Weltbürger.* Stuttgart:
Kohlhammer.

Martin, B. R. and other Editors of *Research Policy SPRU – Science and Technology
Policy Research.* (2007). Keeping plagiarism at bay – A salutary tale. [Editorial].
In: *Research Policy*, Vol. 36, pp. 905–911.

Mathiopoulos, M. (1987). *Amerika: das Experiment des Fortschritts. Ein Vergleich
des politischen Denkens in Europa und in den USA.* Paderborn: Schöningh.

Mathiopoulos, M. (1989). *History and Progress: In Search of the European and
American Mind.* New York: Praeger.

Medick, V. (2011). Prinz von Preußen zu Plagiatsaffäre: "Entschuldigen
kann man so etwas nicht" In: *Spiegel-Online*, 25 February. [Web
page]. http://www.spiegel.de/politik/deutschland/prinz-von-preussen-zu-
plagiatsaffaere-entschuldigen-kann-man-so-etwas-nicht-a-747175.html cited 22
August 2013.

Meichsner. I. (1990). Hübsch geklaut: Kölner Professorin muß um ihren Doktortitel
fürchten. In: *Die Zeit*, Vol. 45, No. 44, 26 October, p. 22. Available at http://
pdfarchiv.zeit.de/1990/44/huebsch-geklaut.pdf cited 16 August 2013.

MMDoku Wiki. (n.d.). *Dokumentation zur Kritik an der Dissertation von Margarita Mathiopoulos.* [Web site]. http://de.mmdoku.wikia.com/wiki/MMDoku_
Wiki cited 22 August 2013.

Mommsen, T. (1876). Die Promotionsreform. In: *Preußische Jahrbücher*, Vol. 37, April, pp. 335–352. Digital version of a reprint available at http://archive.org/stream/DieReformDerDoctorpromotion#page/n39/mode/2up cited 21 July 2013.

Moser, H. (1968). Notiz. In: *Zeitschrift f. dt. Philologie*, Vol. 87, No. 1, pp. 312–316.

Münch, I. von. (2006). *Promotion*. (3rd ed.) Tübingen: Mohr Siebeck.

Münch, I. von. (2012). *Gute Wissenschaft*. Berlin: Duncker & Humblot.

Nix, U. M. (1963). *Der mystische Wortschatz Meister Eckharts im Lichte der energetischen Sprachbetrachtung*. Düsseldorf: Schwann.

Nix, U. M. & Öchslin, R., (Eds.) (1960). Meister Eckhart der Prediger. Festschrift zum Eckhart-Gedenkjahr. Hrsg. im Auftrag der Dominikaner-Provinz Teutonia. Freiburg, Basel, Wien: Herder.

Oberbreyer, M. (1878). *Die Reform der Doctorpromotion: Statistische Beiträge*. Eisenach:Bacmeister. Digital version available at http://archive.org/stream/DieReformDerDoctorpromotion#page/n5/mode/2up cited 10 July 2013.

Pädagogischer Verlag Schwann. (1968). Mitteilung des Verlags. [Editorial]. In: *Wirkendes Wort*, Vol. 18, p. 288.

Piper, M. (2013). Briefe an Prof. Dr. Ernst-Ludwig Winnacker und Prof. Dr. Kurt Biedenkopf. Available at http://www.uni-duesseldorf.de/home/universitaet/weiterfuehrend/pressebereich/hhu-informationsservice-promotionspruefungsverfahren-prof-dr-schavan/briefe-an-prof-dr-ernst-ludwig-winnacker-und-prof-dr-kurt-biedenkopf.html cited 2 July 2013.

PlagDoc & Kotynek, M. (2012). *Schwarmgedanken*. [Web page], 8 June. http://de.guttenplag.wikia.com/wiki/Schwarmgedanken with an English translation by WiseWoman, *Reflections on a Swarm* to be found at http://de.guttenplag.wikia.com/wiki/Reflections_on_a_Swarm cited 15 August 2013.

Plagin Hood. (2013). *Wie entstand das VroniPlag Wiki Projekt?* [Web page]. http://de.vroniplag.wikia.com/wiki/Benutzer:Plagin_Hood/Projektentstehung cited 15 August 2013.

Preuß, R. & Schultz, T. (2011a). Verteidigungsfall. In: *Süddeutsche Zeitung*, 16 February, p. 2.

Preuß, R. & Schultz, T. (2011b). Verteidigungsfall. In: *Süddeutsche Zeitung*, 16 February, [Online newspaper]. http://www.sueddeutsche.de/politik/plagiatsvorwurf-gegen-verteidigungsminister-guttenberg-soll-bei-doktorarbeit-abgeschrieben-haben-1.1060774 cited 15 August 2013.

Preuß, R. & Schultz, T. (2011c). *Guttenbergs Fall: Der Skandal und seine Folgen für Politik und Gesellschaft*. Gütersloh: Gütersloher Verlagshaus.

Preußen, F. W., Prinz von. (1972). *Bismarcks Reichsgründung und das Ausland. Göttingen*. Hannover: Göttinger Verlagsanstalt.

Preußen, F. W. Prinz von. (1984). *Die Hohenzollern und der Nationalsozialismus*. [PhD thesis, University of Munich].

Quint, J. (1964). Besprechungen: Udo Nix, Der mystische Wortschatz Meister Eckharts im Lichte der energetischen Sprachbetrachtung. [Book review]. In: *Beiträge zur Geschichte der deutschen Sprache und Literatur*, Vol. 86, pp. 401–411.

Rasche, U. (2007). Geschichte der Promotion in absentia. Eine Studie zum Modernisierungsprozess der deutschen Universitäten im 18. und 19. Jahrhundert. In: R. D. Schwinges (Ed.) *Examen, Titel, Promotionen – Akademisches und staatliches Qualifikationswesen vom 13. bis zum 21. Jahrhundert.* Basel: Schwabe, pp. 275–352.

Rasche, U. (2013). Mommsen, Marx und May: Der Doktorhandel der deutschen Universitäten im 19. Jahrhundert und was wir daraus lernen sollten. In: *Forschung & Lehre*, No. 3, pp. 196–199. Available at http://www.forschung-und-lehre.de/wordpress/?p=12892 cited 10 May 2013.

Rehbock, T. (1907). *Der wirtschaftliche Wert der binnenländischen Wasserkräfte unter besonderer Berücksichtigung des Grossherzogtums Baden: Festrede bei dem feierlichen Akte des Rektoratswechsels an der Großherzoglich Technischen Hochschule Fridericiana zu Karlsruhe am 30. November 1907.* Karlsruhe: Braun.

Retraction Watch. (2011a). *Ladies and gentlemen, we have an apparent retraction record holder: Joachim Boldt, at 89.* [Blog], 2 March. http://retractionwatch.wordpress.com/2011/03/02/ladies-and-gentlemen-we-have-an-apparent-retraction-record-holder-joachim-boldt-at-89/ cited 30 June 2013.

Retraction Watch. (2011b). *Bulfone-Paus retraction count grows to 13 with one in Transplantation.* [Blog], 13 December. https://retractionwatch.wordpress.com/2011/12/13/bulfone-paus-retraction-count-grows-to-13-with-one-in-transplantation/ cited 30 June 2013.

Retraction Watch. (2013). *Ulrich Lichtenthaler now up to 12 retractions.* [Blog], 24 May. https://retractionwatch.wordpress.com/2013/05/24/ulrich-lichtenthaler-now-up-to-12-retractions/ cited 30 June 2013.

Rieble, V. (2010). *Das Wissenschaftsplagiat: Vom Versagen eines Systems.* Frankfurt/Main: V. Klostermann.

Rosenfeld, H. (1969). Zur Geschichte von Nachdruck und Plagiat: mit einer chronologischen Bibliographie zum Nachdruck von 1733-1824. In: *Börsenblatt für den deutschen Buchhandel*, Frankfurt Edition, Vol. 25, No. 100, 16 December, pp. 3211–3228.

Ruh, K. (1964). Der mystische Wortschatz Meister Eckharts im Lichte der energetischen Sprachbetrachtung by Udo Nix. [Book review]. In: *Zeitschrift für deutsches Altertum und deutsche Literatur*, Vol. 93, No. 1, pp. 31–32.

Sattler, S., Graeff, P., & Willen, S. (2013). Explaining the Decision to Plagiarize: An Empirical Test of the Interplay Between Rationality, Norms, and Opportunity, In: *Deviant Behavior*, Vol. 34, No. 6, pp. 444–463. Available at http://dx.doi.org/10.1080/01639625.2012.735909 cited 27 July 2013.

Schaltenbrand, S. (1994). *Alles gestohlen? Vom Plagiat zur Wiederholung.* Berlin: Frieling.

Schavan, A. (1980). *Person und Gewissen: Studien zu Voraussetzungen, Notwendigkeit und Erfordernissen heutiger Gewissensbildung.* Frankfurt (Main): R. G. Fischer.

Schmidt, R. (2012). *schavanplag: Dokumentation mutmaßlicher Plagiate in der Dissertation von Prof. Dr. Annette Schavan.* [Web site]. http://schavanplag.wordpress.com cited 22 August 2013.

Schimmel, R. (2011). *Von der hohen Kunst ein Plagiat zu fertigen: Eine Anleitung in 10 Schritten. Mit Geleitwort von Karl-Theodor zu Guttenberg.* Münster: Lit.

Shell, K. L. (1991). Review, In: *Amerikastudien*, Vol. 36, Nr. 4, pp. 567–569.

Socher, G. O. (1929). *Zu: 38.0.501.29 des Landgericht I, Socher ./. Englisch, Plagiat.* [Electronic edition], Aachen: semantics, 2013. Digital version available at http://visuallibrary.net/urn/urn:nbn:de:s2w-3484 cited 22 July 2013.

Soreth, M. (1991). *Kritische Untersuchung von Elisabeth Strökers Dissertation über Zahl und Raum: nebst einem Anhang zu ihrer Habilitationsschrift.* Köln: P & P Verlag.

Soreth, M. (1996). *Dokumentation zur Kritik an Elisabeth Strökers Dissertation.* 2nd edition. Köln: P & P Verlag.

Der Spiegel. (1973). Still behandelt. Vol. 27, No. 31, 30 July, p. 42. Available at http://www.spiegel.de/spiegel/print/d-41955163.html cited 30 March 2013.

Der Spiegel. (1989). Kern der Leistung. Vol. 43, No. 37, 11 September, pp. 61–62. Available at http://www.spiegel.de/spiegel/print/d-13496867.html cited 19 January 2013.

Spiegel Online. (2011). *Fußnoten-Streit: Dr. Guttenberg nennt Plagiatsvorwürfe abstrus.* [Online media], 16 February. http://www.spiegel.de/politik/deutschland/fussnoten-streit-dr-guttenberg-nennt-plagiatsvorwuerfe-abstrus-a-745919.html cited 15 August 2013.

Spiewak, M. (2011). Flachforscher. In: *Die Zeit*, No. 25, 25 August. Available at http://www.zeit.de/2011/35/Doktorarbeit-Medizin-Forschung. cited 2 July 2013.

Statistisches Bundesamt. (2011). *Promovierende in Deutschland 2010.* [Report]. https://www.destatis.de/DE/Publikationen/Thematisch/BildungForschungKultur/Hochschulen/Promovierende5213104109004.pdf cited 1 January 2013.

Statistisches Bundesamt. (2012). *Bildung und Kultur – Prüfungen an Hochschulen 2011.* Fachserie 11, Reihe 4.2. [Report]. https://www.destatis.de/DE/Publikationen/Thematisch/BildungForschungKultur/Hochschulen/PruefungenHochschulen2110420117004.pdf and https://www.destatis.de/DE/PresseService/Presse/Pressemitteilungen/2012/06/PD12_209_213.html [Press notice]. cited 19 January 2013.

Sticker, G. (1931). Goethes Zahnleiden. In: *Deutsche Allgemeine Zeitung*, Beiblatt. Vol. 70, No. 461, 17 October.

Streuer, O. (2012). Operation Plagiatsvorwurf. In: *Mannheimer Morgen – Morgenweb.* 23 May. Available at http://www.morgenweb.de/mannheim/hochschule/operation-plagiatsvorwurf-1.584308 cited 17 July 2013.

Ströker, E. (1953). *Zahl und Raum: Wissenschaftstheoretische Studien über zwei fundamentale Kategorien der mathematischen Naturwissenschaft mit besonderer Berücksichtigung der Ontologie Nicolai Hartmanns.* [PhD thesis], Bonn.

Ströker, E. (2000). *Im Namen des Wissenschaftsethos. Jahre der Vernichtung einer Hochschullehrerin in Deutschland 1990–1999.* Berlin: Berliner Debatte Wissenschaftsverlag.

Terp, S. (2013). *Keine Titelaberkennung: TU Berlin entzieht Dr. Goldschmidt den Doktortitel nicht.* [Press release], 9/2013, 18 January. Available at http://www.tu-berlin.de/?id=129525 cited 21 July 2013.

Theisohn, P. (2009). *Plagiat: eine unoriginelle Literaturgeschichte.* Stuttgart: Kröner.

Theopold, W. (1964). *Schiller. Sein Leben und die Medizin im 18. Jahrhundert.* Stuttgart: G. Fischer.

Trauner, S. (2013). Kein flächendeckender Plagiats-Check an hessischen Universitäten. In: *Main-Netz*, 14 April. Available at http://www.main-netz.de/nachrichten/regionalenachrichten/hessenr/art11995,2556569 cited 10 July 2013.

Ullrich, H. (2002). Goethes Schädel und Skelett. In: *Anthropologischer Anzeiger*, Vol. 60, Dezember, pp. 341–368.

Ullrich, H. (2006). Goethes Skelett – Goethes Gestalt. In: *Goethe Jahrbuch*, Vol. 123, pp. 167–187. Digital version available at Google Books.

Universität Basel, Oeffentliche Bibliothek. (1913). *Jahresverzeichnis der Schweizerischen Hochschulschriften 1912–1913 / Catalogue des Écrits Academiques Suisses 1912-1913.* Basel: Schweighauserische Buchdruckerei. Available at http://archive.org/stream/jahresverzei191213univuoft#page/n5/mode/2up cited 17 August 2013.

Veil, W. H. (1946). *Goethe als Patient.* Jena: G. Fischer.

Verwaltungsgericht Freiburg/Br. 1. Kammer. (2010). *Urteil VG Freiburg (Breisgau) K 2248/09, Entziehung des Doktorgrads wegen wissenschaftlichen Fehlverhaltens.* [Legal ruling], 22 September. Available at http://www.landesrecht-bw.de/jportal/?quelle=jlink&docid=MWRE100003242 cited 30 July 2013.

Verwaltungsgericht Köln. (2012). *Urteil VG Köln 6 K 2684/12.* [Legal ruling], 6 December. Available at http://www.justiz.nrw.de/nrwe/ovgs/vg_koeln/j2012/6_K_2684_12urteil20121206.html cited 2 July 2013.

Verwaltungsgerichtshof Baden-Württemberg 9. Senat. (2011). *Doktorgrad; Unwürdigkeit zur Führung; Verstöße gegen die Grundsätze guter wissenschaftlicher Praxis. Az: 9 S 2667/10.* [Legal ruling], 14 September. Available at http://www.landesrecht-bw.de/jportal/?quelle=jlink&docid=MWRE110002949 cited 30 July 2013.

Viebig, P. (2013). Volk darf Doktortitel behalten. In: *Nürnberger Zeitung*, 7 February. Available at http://www.nordbayern.de/nuernberger-zeitung/nz-news/volk-darf-doktortitel-behalten-1.2677618 cited 2 July 2013.

VroniPlag Wiki. (2011a). *Eine kritische Auseinandersetzung mit der Dissertation von Prof. Margarita Mathiopoulos: Amerika: das Experiment des Fortschritts. Ein Vergleich des politischen Denkens in Europa und in den USA. Bericht.* [Web page]. http://de.vroniplag.wikia.com/wiki/Mm/Bericht-20110627 cited 22 July 2013.

VroniPlag Wiki. (2011b). *Ut > Befunde > Überblick.* [Web page]. http://de.vroniplag.wikia.com/wiki/Ut/Befunde?file=Ut_Überblick.png cited 22 August 2013.

VroniPlag Wiki. (2012a). *Analyse:As.* [Web page]. http://de.vroniplag.wikia.com/wiki/Analyse:As cited 22 August 2013.

VroniPlag Wiki. (2012b). *Historische Dissertationsplagiate.* [Web page]. http://de.vroniplag.wikia.com/wiki/Forum:Historische_Dissertationsplagiate. cited 22 August 2013.

VroniPlag Wiki. (2013a). *VroniPlag Wiki – kollaborative Plagiatsdokumentation: Kritische Auseinandersetzungen mit Hochschulschriften auf Basis belastbarer Plagiatsfundstellen.* [Web site]. http://de.vroniplag.wikia.com/wiki/Home. cited 22 August 2013,

VroniPlag Wiki. (2013b). *FAQ.* [Web page]. http://de.vroniplag.wikia.com/wiki/VroniPlag_Wiki:FAQ?oldid=134440 cited 3 July 2013.

VroniPlag Wiki. (2013c). *Quellenaufnahme für Gerhart / Hubatsch (1950) in der Dissertation von Fwp.* [Web page]. http://de.vroniplag.wikia.com/wiki/Quelle:Fwp/Gerhardt_Hubatsch_1950 cited 29 July 2013.

VroniPlag Wiki. (2013d). *Übersicht.* [Web page]. http://de.vroniplag.wikia.com/wiki/Übersicht cited 15 August 2013.

VroniPlag Wiki. (2013e). *Statistik.* [Web page]. http://de.vroniplag.wikia.com/wiki/VroniPlag_Wiki:Statistik cited 15 August 2013.

Waldassen, F. (1933). *Handel mit Doktor-Titeln: ein Spaziergang auf dem Jahrmarkt der Eitelkeiten.* Berlin-Karlshorst: Hannibal-Verlag [Pseudonym of Paul Englisch].

Weber-Wulff, D. (2013). Whistleblowing in Germany. In: *Copy, Shake & Paste.* [Blog], 9 July. http://copy-shake-paste.blogspot.de/2013/07/whistleblowing-in-germany.html cited 15 July 2013.

Weihrauch, M., Starte, J., & Papst, R. (2003). Die Medizinische Dissertation – kein Auslaufmodell. Ergebnisse einer Befragung von Promovierenden stehen im Widerspruch zu oft geäußerten Meinungen. In: *Deutsche Medizinische Wochenschrift,* Vol. 128, No. 49, pp. 2583–2587.

Wissenschaftsrat. (2004). *Empfehlungen zu forschungs- und lehrförderlichen Strukturen in der Universitätsmedizin.* Drucksache 5913/04, 30 January. Available at http://www.wissenschaftsrat.de/download/archiv/5913-04.pdf cited 19 January 2013.

Zehnpfennig, B. (1997). Das Experiment einer großräumigen Republik. In: *Frankfurter Allgemeine Zeitung,* No. 276, 27 November, p. 11. Available at http://www.faz.net/aktuell/politik/das-experiment-einer-grossraeumigen-republik-1590883.html cited 17 July 2013.

Zwenger, C. (1865). Bekanntmachung der philosophischen Fakultät zu Marburg. In: *Archiv der Pharmacie,* Vol. 172, No. 2, p. 192. Available at http://books.google.de/books?id=ifE3AAAAMAAJ&pg=RA1-PA192 cited 18 July 2013.

Chapter 4
Plagiarism Detection

The question arises as to how to go about detecting plagiarism when a paper is deemed to be suspicious. Educators would welcome a simple solution, in which all written materials from students would be submitted to some software system, and then speedily returned with the plagiarized portions all neatly marked. However desirable this kind of a litmus test for plagiarism may be, it is not available now and probably never will be. As was discussed in Chap. 2, there are many variations of plagiarism above and beyond simple copying and pasting. Even for simple cases, software is notoriously poor at detecting all but the clearest copies.

The author has tested the effectiveness, usability, and professionalism of such systems many times in the past decade. In particular, tests of plagiarism detection systems were conducted in 2004, 2007, 2008, 2010, and 2013. Specialized tests were done in 2011 (a test of zu Guttenberg's thesis with a few of the systems) and 2012 (collusion detection, including program code plagiarism). The detailed results of these tests are available online (Weber-Wulff n.d.).

This chapter will be concerned with discussing how the software for finding sources that match a portion of the text in the suspicious thesis works, how to go about searching for plagiarism sources without using such software, a discussion of the methodologies that are used by contributors to the VroniPlag Wiki, and the special case of detecting collusion, which is a situation in which multiple students submit the same or slightly altered versions of a text for a grade.

4.1 Text Matching Software

There are countless systems available that promote themselves as plagiarism detection systems, either systems that must be purchased or ones that can be used online for free. However, it is important to realize they do not determine plagiarism per se, but they flag more or less matching text that must then be evaluated by a teacher before a decision can be taken as to whether or not it is, indeed, plagiarism. The systems use different methods of identifying text matches. The ways in which such systems fit into a university workflow are also quite varied, and as a rule their effectiveness is much less than is generally assumed. This section will attempt to give just a basic overview of how they function. Details on the algorithms used are beyond the scope of this book. A more detailed discussion of technological questions can be found at Mozgovoy, Kakkonen, & Cosma (2010). Further papers describing various aspects of automatic plagiarism detection are Hage, Rademaker, & van Vugt (2011) and especially Lukashenko, Graudina, & Grundspenkis (2007).

4.1.1 System Functionality

Most companies are understandably reluctant to disclose how exactly they go about detecting text similarity. There seem to be two major strategies, however, for finding possible matches for a given text. One involves searching a database that is under the control of the company. The other uses the results of traditional search engines in order to find candidate sources. There is also some experimental work being done in analyzing a text for stylistic inconsistencies.

4.1.1.1 Databases

A text similarity detection system that uses a database takes papers submitted to it and checks them against the database, looking for possible sources. Since it is quite advantageous for companies to compile such collections of written work, they are often quite keen to store all papers submitted so that they can be used for checking future papers. Using a database also means that certain kinds of pattern matching algorithms are easier to implement.

The company iParadigms, for example, which markets *Turnitin* to universities and *iThenticate* to companies, reported that it receives up to 500,000 papers submitted daily to their system by students, and that they processed over 80 million papers in 2012 (iParadigms 2013b). The company keeps three databases, storing more than 24 billion web pages that they have accessed through "crawling" the Internet by following as many links as possible. According to iParadigms, over 300 million student papers that were submitted to the system are also in the database, as well as over 120 million academic articles from over 110,000 journals, periodicals, and books that have been contributed to the system as part of an agreement called *CrossCheck* that academic publishers are using to try and detect duplicate papers before publication. In this scheme, all published papers are added to the database, so that newly submitted journal articles can be checked for plagiarism before being published.

When storing student papers in a database, the question of copyright does need to be considered since the students do hold copyright on all original material they produce themselves. In a 2008 suit filed in an Alexandria, Virginia court (*Vanderhye ex rel. A.V. et al. v. iParadigms, LLC* 2008) in the USA by the parents of four high school students, it was alleged that *Turnitin* was violating the students' copyright by storing a copy of their original papers unasked. The court ruled that iParadigms' "use of Plaintiffs' works has caused no harm to the market value of those works" (2008, p. 16) In fact, it finds what iParadigms does with the papers to be "highly transformative and highly beneficial to the public through educational institutions using Turnitin." (2008, p. 16) The students filed an appeal in 2008, citing the ease of thus obtaining copies of others' works without permission and FERPA[1] rules. The Fourth Circuit Court of Appeals, however, rejected most of their appeal (*A. V. et al. v. iParadigms, LLC* 2009).

In Europe, however, the EU copyright laws are more focused on the rights of the authors. In particular, German copyright law (§ 16 UrhG) gives the author the absolute right to determine who is able to exercise which usage rights for their texts. An expertise obtained from the Intellectual Property Rights Helpdesk of the EU details that students' papers may only be stored in the database of a commercial company with the express permission of the student (IPR Helpdesk 2004).

Universities should go to the trouble of obtaining explicit permission from their students, but should not penalize students who are not willing to donate their papers to furthering the goals of a private company. Instead, teachers could be advised to pay particular attention while reading such a paper to be on the alert for possible plagiarisms. By and large, it is often faster and more effective (see Sect. 4.2) to use a search engine instead of specialized software.

It should also be made explicitly clear to both teachers and students when the company is keeping a paper in a database. In the process of a test of the systems, conducted at the HTW Berlin in 2013, it was found that in addition to *Turnitin*, the Dutch system *Ephorus*, the Swedish system *Urkund*, and the French system *Compilatio* also stored papers submitted. Test cases from previous tests have even been found, although every attempt was made to not store the papers and the companies were explicitly requested to delete any papers submitted in the course of the test from their databases. Systems such as *Viper* and *Turnitin* also give themselves irrevocable rights to do whatever they please with papers that are uploaded. This is a portion of the *Turnitin* end user license agreement (EULA) that was displayed in April of 2011 when registering for use of the system (iParadigms 2011):

> With regard to papers submitted to the Site, You hereby grant iParadigms a non-exclusive, royalty-free, perpetual, world-wide, irrevocable license to reproduce, transmit, display, disclose, archive and otherwise use in connection with its Services any paper You submit to the Site whether or not originally submitted in connection with a specific class. This license shall survive the termination of the User Agreement. Any cessation of use of the Site shall not result in the termination of any license You grant herein to iParadigms.

It has now been amended online to include affiliates, but states in a roundabout way that they will not use the content of the papers (iParadigms 2013a):

> If You submit a paper or other content in connection with the Services, You hereby grant to iParadigms, its affiliates, vendors, service providers, and licensors a non-exclusive, royalty-free, perpetual, worldwide, irrevocable license to use such papers, as well as feedback and

---

[1] Family Educational Rights and Privacy Act of 1974, giving students access to their educational records and control over their disclosure.

results, for the limited purposes of a) providing the Services, and b) for improving the
quality of the Services generally. [...] The licenses shall survive the termination of the User
Agreement. Any cessation of the use of the Site or Services shall not result in the termination
or expiration of this license. [...] This Communications license does not include any right
of iParadigms to use ideas set forth in papers submitted to the site, which is covered by the
license above.

Still, for universities that cooperate with private companies for final theses, especially in technical fields, this will pose a thorny problem. The standard non-
disclosure agreements about the written expression of the work done by the student
would not be compatible with the license agreement above, even in its amended
form.

One system, Viper, even makes it explicitly clear on their web site that they will
be including the papers that are uploaded to their site for plagiarism checking in
their paper mill database (All Answers Ltd. 2012):

> When you scan an essay, we'll take your essay and add it to our database so that future scans
> that you or other people make can be compared to it. Nobody has access to this database
> and if part of your essay matches another essay, other people cannot see your work - they
> only see a percentage match.

> Aside from that, 9 months after your scan, we will automatically add it to our student
> database and it will be published on one of our study sites to allow other students to use
> it as an example of how to write a good essay.

> Most students don't mind helping other students by offering their essay/work as an example
> - they're finished with the essay and it's of no value to them.

Perhaps not surprisingly, the email address that was used for the test of the Viper
system is now regularly receiving spam suggesting the use of that paper mill or their
custom term paper writing service.

4.1.1.2 Local vs. Web-based

The databases mentioned in the last section are normally located at the site of the
company offering the software system. This is problematic, as mentioned, in cases
of non-disclosure agreements. It also means that the texts submitted are now outside
of the control of the organization that is examining the document. Additionally, any
laws governing the data will be the ones applicable to the place in which the server
is located.

Universities would often prefer to be able to keep the database of the texts stored
on a server that is within their own university system in order to comply with local
information technology security guidelines. A few systems such as *BOSS/Sherlock*
advertise that they permit a local database to be set up. Attempts by the author to set
up such a system in 2012 were not successful, however. It appears that this system
has not been updated since 2003.

It is understandable that companies want to have as many papers in their own
databases as possible, but this is an area in which universities should be more demanding. Some systems, such as *Ephorus*, offer "pools" of papers that a university

can join, and they can set up their own pool. The papers are still kept on *Ephorus'* servers, but the company states that only the university that owns the pool may search for matching papers within the pool. A region such as Berlin, for example, could set up a pool for Berlin schools in order to thwart the game of turning in the same paper or thesis at different schools by different students. However, since the databases are under the control of the company, it is still possible for programming errors to expose the theses to the outside, as was seen happening with *Compilatio* during the test of 2013 (Weber-Wulff, Möller, Touras, & Zincke 2013).

Many teachers would prefer to have a local searching tool, that is, one that would be installed on the teacher's computer only, perhaps just keeping a local database for the teacher. There are a few systems available that offer this, but they sometimes have problems such as installing viruses, or not being reachable for support,[2] or being rather unstable. The German system *PlagiarismFinder* had the excellent idea of selling a memory stick with their software installed. Especially high school teachers are quite interested in a system that can be passed around and used on different computers. However, this system did not find much in the way of plagiarism, as tests in 2010 and in 2013 have demonstrated (Weber-Wulff & Köhler 2010a; Weber-Wulff et al. 2013).

4.1.1.3 Intra-Corpal vs. Extra-Corpal Searching

A very important difference between plagiarism detection systems, independent of whether or not they use a database, is the question of intra-corpal vs. extra-corpal searching for text parallels. An intra-corpal search is done when there is a closed set of documents given that are to be examined, e.g. all of the student submissions for one class are to be checked for text parallels. This happens when two or more students in a class decide to work together and hand in the same or similar papers, although they were instructed to work alone. This is known as *collusion*. Searching for collusion within a closed set is relatively easy as the system only has to check each paper against every other one. Section 4.5 will discuss a specific test of collusion detection systems.

In an extra-corpal search, the sources for the text parallels, if they even exist, are not part of a closed set of documents, but to be found in the seemingly infinite expanses of the Internet. It is also not even clear if sources exist. If there are sources, they could be part of the open, accessible web, but they could also be hidden behind a so-called paywall, which makes a user pay to access the source. Or they could be stored in databases that are not indexed by search engines. There could also be sources in books that are not (yet) digitized. This makes it difficult to find all the sources, or even one possible source, for a text parallel.

---

[2] Obtaining promised refunds can also be difficult. The "free" trial of *Eve 2* consisted of paying up front and then supposedly getting the money back no-questions-asked if requested within 15 days. An attempt in 2008 to obtain a refund was not successful; no one has ever been available at the company since. The support email address bounces, but a call center answers the telephone and is always willing to take credit card numbers. The web site is still online as of 2013.

Depending on whether the detection is to be intra-corpal or extra-corpal, there are different algorithms that can be used. All have to deal with the complexities of size. While it is computationally feasible to compare 50 student papers with each other (although as Lancaster (2003) has demonstrated, it is not trivial to visualize the results), as soon as the number rises much beyond 100 the problem quickly becomes very complex. If there are n papers to compare with each other, then a collusion investigation will need to conduct $n(n-1)/2$ comparisons. For 100 papers this would be 4,950 comparisons. If each one only took a second, it would still take almost one and a half hours to complete the task.

A direct comparison between all possible documents is impossible for an extra-corpal search. Instead, the systems attempt to first identify candidate sources, and then look more closely at these candidates to see if they match the text at hand. If the candidate pool is too small, a source might not be found. If the pool is too large, it will take a long time to examine. This is why most systems will only choose a sample of the papers, especially for large documents, and look for possible matches only on the sample.

4.1.1.4 Using Search Engines

A popular method for finding candidate sources for plagiarism is to use the search engines available. Most search engines offer an API, an Application Programming Interface, that permits a program to send a query to the search engine and obtain back information on possible matches for the query, much like using a search engine through a web interface. However, excessive use of this is quite expensive. Google (2013) quotes some prices on its web site:

> Any usage beyond the free usage quota [100 free queries per day] will fail if you are not signed up for billing. Once you have enabled billing, you will continue to receive 100 free queries per day. However, you will be billed for all additional requests at the rate of $5 per 1000 queries, for up to 10,000 queries per day.

In addition, no results are returned that include Google Books, as there is a specific licensing agreement with book publishers on the use of the snippets from the books that have been scanned in. So for a system to integrate a Google search into its algorithms, this will either mean that only a few samples of the text can be investigated, or that a thorough check will be very expensive – or else take very long to finish, as only 100 free queries are permitted per day. The companies offering search engines can, of course, have forged special deals with the software companies for bulk purchase of queries.

Some plagiarism detection software only does a superficial search, setting up a web page link that includes the search terms so that when it is clicked on, it simulates a web-based search request. The user then lands on the normal results page of the search engine. For this one does not need a software system – it would be sufficient and faster to search oneself. Techniques for identifying good search phrases will be discussed in Sect. 4.2.

4.1.1.5 Techniques for Finding Text Parallels

In searching for source candidates, either in a database or with a search engine, the systems also use a number of different methods. Lancaster & Culwin (2005) give a short synopsis of the different techniques that were identified by Lancaster in his dissertation (2003).

For each document, a selection of numbers is calculated, called metrics. For example, if program code is being analyzed, one can count the number of different variables, the number of different method names, or the absolute number of variable and method usages. Halstead (1977) introduced this type of metric in an attempt to characterize the size of computer programs. It turns out to also be a good method for finding plagiarisms in student computer programs, as just counting the number of occurrences ignores the actual names of the variables and methods, which could have been renamed by a plagiarist.

Other kinds of metrics count attributes such as the average number of words per sentence or the ten most frequent words in a document. Some look at structures between two documents such as common runs of characters or words (often called n-grams) or the number of capitalized words found in common. Many of the metrics are statistical in nature. Whichever metrics a system uses, they are collected up in a particular order and examined to see how close these metrics are to each other for the two documents they represent.

This determination of "closeness" is often referred to as a similarity score and is generally normalized to a percentage so that identical copies are 100% similar. Some of the metrics may be given more weight in the calculation than others. Formatting issues, for example, hyphenation, may cause documents that are, indeed the same to not be given a score of 100%.

For systems that use a database for storing copies of potential sources, the collection of metrics is usually stored in the database as well. These attributes are often referred to as the "fingerprint" of the document. It is, however, entirely possible that two documents can have the same fingerprint even though they are different, so the term fingerprint is somewhat misleading here. This can lead to a situation in which the plagiarism detection software announces a match when actually there is no commonality at all.

Instead of calculating a similarity score, some systems look for the longest sequence of characters in two texts that are identical. Since this would not be feasible for the entire document, these systems will generally just take a sample out of the document to be checked. The sample chosen might be one or more blocks of text, and it may be chosen at random. Running the same text through the same system more than once can thus produce wildly different results for each attempt. Another algorithm observed during testing chose seven words running, then skipped six words and took the next seven words. Unfortunately, once it found matches, it never attempted to extend the edges of the matches, so it announced a much lower amount of plagiarism than was actually contained in the text.

The systems that use word counting for selecting the samples often run into problems with long names of institutions (such as "der Rechts- und Sozialwissenschaftlichen Fakultät der Universität Bayreuth", which was flagged by *Turnitin* as plagiarism), or legal texts ("§ 3 (4) 1 UrhG" counts as five words), or with non-Latin alphabets. A Japanese sentence, for example, looks just like one long word to such a system.

Many of these systems do not work well for texts that are even slightly changed. For example, if the source text has been changed from "Iceland and Greenland" in the original to "Greenland and Iceland" in the copy, sequence comparing systems often do not see that these are very similar. This is because of the combinatorial explosion in number of possible variations on the word order. It is impossible to try all possible combinations using a software system, but human beings, who have highly developed pattern matching facilities, can easily spot the similarity.

Other obstacles to finding matches include the use of diacritical marks, which often have alternate representations and can confuse systems, as well as writing out numbers in the copy, as shown in Fig. 4.1. In this fragment, a number and a symbol were written out and an abbreviation expanded.

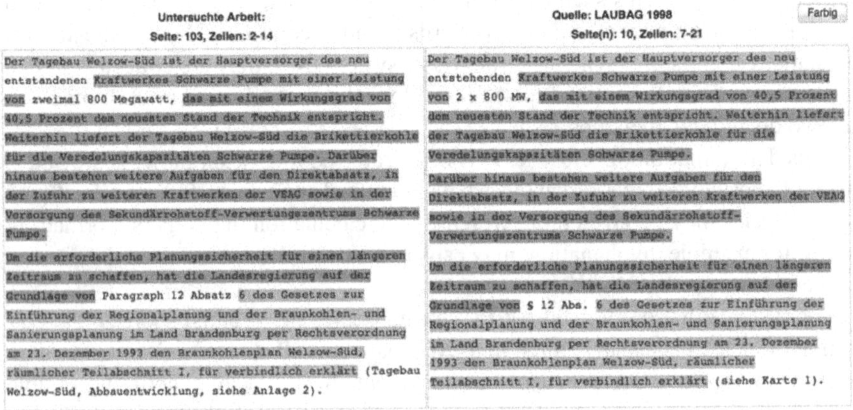

Fig. 4.1 A fragment from the thesis *Dd* (VroniPlag Wiki 2011a)

Writing out a symbol or number like this can confuse a plagiarism detection system because it only looks at the exact characters, not the meaning. Once a match is made, some systems do attempt to extend the match by attempting to add text to the beginning and the end of a match, or by hopping over a word and trying to restart the matching algorithm.

4.1.2 Testing the Effectiveness

Many people are convinced that plagiarism detection systems are effective, just be-cause they do find sources for some of the material that they investigate. This is, however, not evidence of exactly how effective a system really is.

In order to test the effectiveness of plagiarism detection systems, a collection of plagiarism test cases has been developed at the HTW Berlin since 2004. There are now 61 short texts and three longer ones in this collection, in English, German, Japanese, and Hebrew. The texts are either copy & paste plagiarisms; translations from English, French, and Norwegian to German and from German and French to English; shake & paste collages; disguised material; copies from copyright-free material available on Google Books; clause quilts; or structural plagiarisms. There are also a number of original texts, in order to see how the systems react to such material. Two of the original texts were also added to the Wikipedia, in order to see how well the systems work at finding Wikipedia content.

The most recent test of plagiarism detection systems was completed in August 2013 using 33 test cases to see how effective 16 systems were in identifying known plagiarism in short essays and in not registering plagiarism in original papers. Al-though the test was conducted with a point-ranking prepared in advance, it soon turned out that the points were rather meaningless. The percentages of plagiarism reported by the systems were so varied, and many would report the correct source mixed in with dozens of irrelevant links. It did not make sense to give a system more points for reporting a higher percentage and the correct link as one of as many as 50 sources than one which only reported the correct source but with a much smaller, incorrect percentage.

The tests were conducted by students from different areas of study[3] who at the end of every test evaluated the usability of the system, first as a general judgment based on the use of the system in a university environment and then systematically using a checklist. There was a wide variance between their perceptions of the ease of use or utility of the various systems, but they were eventually able to reach a consensus for each system. In addition, systems that were easy to use often were not very good at identifying possible sources, while good systems were quite difficult to use.

The systems *Urkund*, *Turnitin*, and *Copyscape* were found to be partially use-ful as tools for identifying candidate sources for plagiarism that can be submitted digitally. However, each of the systems has its own usability issues. Other systems would occasionally identify a source properly, but they were often not consistent in their performance.

---

[3] I wish to thank Christopher Möller and Matthias Zarzecki, bachelor students in International Media and Computing at the HTW Berlin; Jannis Touras, doctoral student in Philosophy at the HU Berlin; Elin Zincke, master's student in Library and Information Science at the HU Berlin for their tireless work and the many fruitful discussions we had during the course of the tests. Matthias Zarzecki produced the new test cases for this test, while the other three students conducted the tests under the supervision of the author.

Not all of the companies selling the software were willing to give out price information as that is often individually worked out with each institution. Some of the systems can be attached to popular learning management systems for better integration in the university process. Detailed results of the test are beyond the scope of this book, but can be found online at the HTW Berlin Plagiarism Portal (Weber-Wulff n.d.).

None of the systems can be recommended for general use in scanning all papers handed in to a university because all systems suffer both from false positives (non-plagiarized text flagged as plagiarism) and false negatives (plagiarisms not found). In particular, since the systems do not test the entire text but only samples, when they are used on larger texts such as final theses, they may miss a good bit of plagiarism. The systems can, of course, be used for large, first-year courses in order to demonstrate that it is possible to pick out plagiarism, but this also incurs the danger of students manipulating their texts so that they "pass" the software without flagging plagiarism. *iThenticate*, for example, a variant of *Turnitin*, offers authors the possibility of purchasing an individual test and then re-checking five revisions without additional cost. This focuses on writing to the machine, not independent writing.

Text-matching software should be seen as just one additional tool in the toolset that can be used when there is a suspicion of plagiarism but nothing has turned up with a search engine.

4.2 Just use Google

Many people are convinced that it is difficult to find plagiarism without using some sort of plagiarism detection software. If one has a suspicion, however, it is not very difficult to find a source, although it is not trivial to find all of the sources. This is, however, often not necessary, as it is sufficient at university to demonstrate the existence of some plagiarism in order for a paper to be declared failed, instead of needing to determine all such passages.

4.2.1 Read and Google

The first step in finding plagiarism is a careful read of the paper in question. This is a point that is contentious as many teachers would prefer to have the plagiarized papers weeded out before they begin reading so that they can concentrate on only giving feedback to the honest authors. But since software cannot determine if a paper contains plagiarism or not for certain, teachers still must examine each paper submitted.

While reading a student's submission, the educator needs to be acutely aware of the style – is this a typical student's style, or is it somehow more polished than one would expect from a student? Is the vocabulary level too high or is the student using unusual phrases? Interesting errors such as misspellings of proper names or style shifts or even formatting shifts can be indications of plagiarism. Are the sources all relatively old? For example, if the paper was submitted in 2012, but the most recent source is from 2007, this may be an indication that an old paper has been recycled. Searching for a source while reading can interrupt one's concentration, however, so it is generally a good idea to just make a note of such anomalies on the margins of the thesis while reading and to undertake an online search later on.

In order to search for possible sources, one does not need to have the entire thesis digitalized.

It is often sufficient to take **three to five words** from a paragraph and submit them to a search engine.

Good candidates are words that are not very common, and nouns tend to be good selections. For example, using just the terms "spectre misconduct politicians" (without quotation marks), Google returns the author's BBC article on plagiarism in Europe as one of the first entries. Using good short phrases, e.g. "papers pounced" and "odd footnote" (each phrase now enclosed in quotation marks), there are only a few results returned: the article and some blog posts quoting these phrases.

This seems quite illogical to many people as they are used to typing just one word into a search engine and having to contend with millions of hits. However, with each word that is added to a search, the field is narrowed down considerably. And even if a student switches a few words around – something that might fool a plagiarism detection software system – the search engine will not be irritated, as these algorithms look for the words close to each other, no matter what the order of the words is.

The German-language e-learning unit *Fremde Federn Finden* (Weber-Wulff 2007) offers exercises in Chapter 6, ten of which are in English. These exercises are the short essays that were used as test cases for testing plagiarism detection software. They are useful for training teachers to spot good terms for searching for a source using a search engine. There are answers linked for every essay with possible search terms, so that even if a search is unsuccessful, the solution can be consulted and the appropriate search terms tried out.

4.2.2 Google Books

Google has been digitizing books and making them full-text searchable since 2004. Google does not publish numbers, but as of March 2012, according to an article published in the *Chronicle of Higher Education* (Howard 2012), the number of scanned books was over 20 million.

One can search Google Books just like searching the web – put a keyword or a phrase into the search box, and see what shows up. The presentation of the results, however, is rather complicated and has to do with copyright laws. And since these differ from country to country, the same search can lead to different results in different countries.

Depending on the copyright situation, either a preview of the book can be seen, e.g. the table of contents and a few selected pages, or the entire text. For works in the public domain, a PDF download is offered.

For the rest, Google just offers a *snippet*, two or three lines of the text with the search terms highlighted. Sometimes this is enough to determine that a book needs to be obtained for further investigation, other times it is just maddening enough to not quite have enough to see if the book could be a source. There are usually links offered for obtaining the book from a library or for purchasing it, but they can be unreliable as they do not include all possible avenues for finding a physical copy of a book.

4.2.3 Google Scholar

Google Scholar is a specialized database that offers a full-text index of some scholarly literature. Since 2004 this service has attempted to find and index peer-reviewed online journals from various scholarly publishers, as well as books and non-peer-reviewed material. There are a number of other, similar databases, either offered by the publishers themselves or at sites such as Thomson ISI's *Web of Science* or *CiteSeerX*.

An advantage of searching directly on the Google Scholar page is that there is no advertising and there are no result pages that are just cleverly disguised marketing pages. There are a surprising number of academic journals that are indexed here, although Google does not publish a list of the journals indexed. In general, though, the results will be links to pages that are behind a paywall. These are pages set up by the publishers to require payment, often $20 to $30, for access to an article. Logging in from a university account on the university network and searching again through the library system can often access the same papers without cost.

Citations and references that are documented incorrectly (misspellings in author or title, wrong issue or page numbers) in the thesis under investigation can also be used as good search terms here, as this will turn up other papers that use the same, wrong information. Sometimes surprising clusters of identical papers submitted to different conferences by different individuals can be identified in this manner. These kind of conferences can be considered mock conferences, as they accept any paper submitted, as long as the author pays the (often steep) conference fee.

4.2.4 Paper Mills

Educators are often concerned that with the proliferation of paper mills they will not be able to detect papers that were purchased and handed in by a student as their own. This is, indeed, a pressing problem, as custom-written papers produced by a good ghostwriter will in general be plagiarism-free. Even if the students are required to hand in their work in stages, i.e. first an outline, then a rough draft, and then a final draft, complete packages of papers including all of these stages can be purchased for an additional fee.

In order to counteract this, students can be required to note down the call numbers in the local library for all the sources used in their papers. That is something that honest students will have to do anyway, and it would be a good exercise for corner-cutting students. But there is no known means of discovering well-written, purchased papers, except perhaps through the administration of a viva, an oral exam.

Many students who purchase papers, however, are either looking for the cheapest paper possible or do not bother changing anything but the first and last name of the author. Entering the title of a suspicious paper into the search field of a paper mill or just into Google can turn up rewarding results. Educators should not purchase any papers that they find in this manner but should first contact the site owners. There are some sites that will cooperate with teachers. If an email is sent from a university account requesting a copy of a paper that is suspected as a source, it will be promptly sent out, free of charge.

4.3 The Art of Finding Plagiarism

How does one go about finding plagiarism if the document in question is not digitalized? Even if it were, the available plagiarism detection software is not completely reliable, as was seen in the previous sections. This section is devoted to explaining the possibilities for finding and documenting plagiarism without using so-called plagiarism detection software.

4.3.1 Identifying Candidate Sources

Reading the thesis at hand and looking up interesting terms using a search engine, as discussed in Sect. 4.2, is one of the first steps that needs to be taken when investigating a thesis for plagiarism. There are a number of other strategies that can be used for identifying good candidate sources.

Strangely enough, many authors seem to be of the opinion that it is sufficient for them to only name their sources in a footnote – or even just in the reference section – without making clear exactly how much has been taken, often verbatim, from the source. Thus, it can be useful to look through the footnotes and the works cited, even to the extent of obtaining every single work listed as a reference. This method is called *brute force* because every named source is looked at instead of just promising ones. This was the method used for discovering the plagiarism in the dissertation of German education minister Annette Schavan, as well as for a number of the VroniPlag Wiki cases. In the process of obtaining the works listed, it is sometimes possible to identify additional, related works that may also turn out to have been used as sources.

Each and every work named must be obtained and compared with the thesis in question. This has the advantage of turning up even minor phrases copied from one of the sources, but it incurs a steep cost with respect to time and effort because each work must be at least partially digitalized, as described in Sect. 4.3.2, and compared with the thesis. The reference list can also be analyzed in order to see if a particular journal is used disproportionately often. If the author is relying closely on many different papers from just one journal, there may be misappropriated material from there as well.

It can also be interesting to undertake a statistical analysis of the references. Are most (or all) of the sources oldish in relation to the date the thesis was submitted? This could be an indication that it is a (partial) copy of some prior work. Are there papers listed in which the author was a co-author? These should be examined in order to see if there is text recycling between these papers and the work being investigated.

One obstacle to the brute force method is the use by some authors of what could be called *garnish references*. These are references, often taken from other cited works, that are included to make the bibliography look much more thorough, without the author having actually read the reference in question. This sometimes becomes clear when the author transcribes the errors that the original author made in their bibliography, especially volume or page numbers or spelling errors in the title, or even when they cite fictitious references that do not exist. Even though this may make a brute force search more difficult, searching for the erroneous references can indeed turn up a promising source that has this exact same wrong reference citation.

Some authors, as described in the section on pawn sacrifice, will correctly set off and footnote a small quotation but continue on verbatim after the quote without attribution. Others set a footnote at the end of every paragraph, and the paragraph itself is then taken from the work given. Following up on these leads can help one discover the sources for copied material.

4.3.2 Digitalization

Once a potential source has been found, it is now easy to compare the paper with the possible source when both have been digitalized. This is not as much of an obstacle as one might imagine. Many libraries now have high-quality book scanners available. These are scanners that make it easy to scan directly from a printed source. Typical examples of these are the *BookEye* or *Zeutschel* scanners, and there is even an active do-it-yourself book scanner community.[4] A book is laid flat on a cradle, open to the pages that are to be scanned. There is a camera and light mounted above the cradle. Both pages can be photographed at the same time, or the system can take a picture of each page individually. There are even systems that recognize the fingers and thumbs holding the pages down so that they can be automatically erased. Other systems have a glass plate that can be used to keep the book flat. After turning the page, the next photo is made until all of the interesting pages have been copied.

Most systems will store the results on a USB stick, usually in PDF or TIFF format. If possible, the TIFF format is better. With a little bit of practice, one can scan about 200 to 300 pages in an hour, depending on the system. Do note that legally the scan is only for private or research use, so it may not be publicly posted on the Internet. It is important that the scan be done in gray scale, not color, and in as high a quality as possible, at least 300 dpi, so that the error rate on character recognition is minimized. If pages are browned, it may be necessary to digitally enhance them after scanning.

In many libraries in Germany, a few cents need to be paid per scan, as the libraries must pay a fee to the German intellectual property rights organizations such as the *VG Wort* for every page copied or scanned. One can also use a smaller home scanner, as some of them can produce good results. Care has to be taken that the pages are properly squared off, or each of the pages will need to be attended to during text recognition. Some problems also arise if pages are larger than the bed of the scanner. Once a scan for each page exists, they must be joined together to form one PDF file. Some home scanner software can deal with this, the professional systems in libraries will generally take care of this step without asking, and there are freeware systems available.

Now optical character recognition (OCR) needs to be performed on the pages scanned in order to extract the text from the pictures produced by the scanner. There are quite a number of software systems available for this task. Many of the cheaper or free ones make mistakes while reading that may need to be corrected by hand. *Abbyy Fine Reader*, *Adobe Acrobat Professional*, or *OmniPage* can be used to obtain a computer-readable version of the text. Of the free OCR systems, Google's *Tesseract* is said to be the most reliable one. Some of the systems, for example the *FineReader*, are also able to recognize the text in tables and graphics and align them properly, as well as re-orient pages that slipped while scanning.

FineReader will permit reassembly of the file with the text placed underneath the picture of the page so that it can be copied and includes many controls for fine-tuning the process. In particular, it has dictionaries installed that will help in the recognition process. It then marks all of the places in which it is not sure that it was able to recognize the text correctly, so that a manual check can be made. However, this is generally not necessary if the goal is only to get a first impression of possible text parallels. If a letter 'c' ends up being recognized as an 'a', it is not a big problem, as this word will not be marked by the comparison program described in Sect. 4.3.3.

---

[4] There are a number of web sites with construction plans for do-it-yourself scanners, for example (DIY Book Scanner n.d.).

The human eye quickly sees that they are the same, however, and can judge whether it is necessary to fix or not.

The files produced by scanning and OCRing tend to be quite large, so they far exceed the size of files that can comfortably be mailed to oneself or others. However, quite a number of services exist online that are useful for transferring large files, for example, *Dropbox*.[5] Even Google now offers a few gigabytes of data storage for free that can be used for transferring such a file, removing it after it has been transferred.

4.3.3 Comparing

Once a potential source has been found, there is an open source program called *SIM* (Grune & Huntjens 1989) that can compare texts, returning a list of positions within the texts that match identically. Since the program is also able to compare a number of texts against each other, or even compare a text with itself, it is also quite useful for cases in which there are multiple sources for one portion of the assumed plagiarism, for comparing papers against each other to discover textual collaboration, and for finding text reuse within a text. Apparently, some authors writing with computers regularly copy blocks of text around and are unaware that they have already used this exact same wording somewhere else in the text. In one thesis that was looked at by VroniPlag Wiki, one block of text was used six times in the thesis. If there had been careful editing of the manuscript by a third party before publication, this would hopefully not have made its way into print.

SIM only works on text files, so the text needs to be extracted from a PDF before the comparison can begin. It does not have a nice input interface, but must be used on the console level, and its output can be quite difficult to read and interpret. It basically just produces a text file listing where a text parallel begins and where it ends. The original program has many parameters that can be set in order to control the minimum run size or the threshold level of percentages to show. Here are two examples of using it:

```
sim_text -o sim_r05_t1_t2.txt -s -d -r5 t1.txt t2.txt
```

compares the file t1.txt with the file t2.txt using a run-length of 5 and outputting in a so-called diff-format to the file sim_r05_t1_t2.txt.

```
sim_text -o sim_self_r07_t1.txt -d -r7 t1.txt
```

compares the file t1.txt with itself, using a run-length of 7 and storing the results in the file sim_self_r07_t1.txt. It can also be set up to recursively look at all sub-directories encountered.

The output information produced by *SIM* can, however, easily be used by other programs to provide better interfaces, and so the *SIM* program was re-implemented

---

[5] In the aftermath of the governmental eavesdropping scandals of 2013, one of the self-hosted examples at Peng (2013) might be considered a better option.

by a VroniPlag Wiki programmer in order to be run in a browser (VroniPlag Wiki n.d.). A suspicious document is copied into the left-hand box and a potential source document into the right hand box. If there are text parallels, they will be marked in the same color in both boxes. This makes it easier to see both the parallels and the changed portions of a text. There is often more than one fragment taken from a particular source, so once one text parallel has been found, checking the entire document against the source can yield additional text parallels. As needed, the user can adjust a parameter defining the minimal number of adjacent words that must be identical in order for the coloring to begin. However, for technical reasons, the self-comparison does not work with this implementation of the *SIM* algorithm.

The coloring of identical text has turned out to be an extremely effective aid in making the extent of the text parallel clear. Not only can the eye quickly see the matching segments, but it is possible to rapidly recognize that the non-marked portions are in essence saying the same thing, or that they have been slightly rewritten, or that they are the result of simple OCR errors. This side-by-side documentation is also good for preparing reports for an examination board or an ombud for good academic practice, as the extent of text copying, at least, is immediately and easily understood.

4.3.4 Using Plagiarism Detection Software

If all else fails, plagiarism detection software can of course be used in order to generate leads on possible sources. However, some of the software is quite expensive and some of the free ones have undesired side effects. Systems have been found that install viruses on the user's machine, and some ghostwriting or search engine optimization companies offer such tools for use, perhaps in order to obtain fresh material for setting up linking material for so-called search engine optimization, or to gather addresses of potential customers for ghostwriting services. One should always read the terms of use closely, as many systems wish to retain copies of the material submitted for comparison with future theses.

Legally, this is something that only the author of the text can assent to in Europe, and the situation is different to that in the USA. Many systems hide the option for not storing the text, or they use terminology to describe what they are doing that does not immediately make it clear that the text will be stored. One system asks if you want to keep the file in your *pool*, another for your *filing system*. They give the impression that this is your private area, so that when you delete the file, it is gone. This is, as our tests have shown, not the case for many different systems. All have apologized for the accidents when we pointed out that they were keeping copies of the files illegally, and assured us that they would delete the files promptly.

In particular, plagiarism detection software systems produce numerous false negatives and false positives and both cause problems for the academic system. Just because a software system does not identify any sources for a paper or thesis does *not* mean that it is free of plagiarism. It only means that nothing could be found by this system. On the other hand, many systems will return quite a large number of "hits" that are either out-and-out wrong, or of very short, minor phrases, or that consist of just bibliographic references, or of references to material that is no longer available at the stored URL. It is a lot of work to wade through all of these false positives looking for a real source. Both of these problems can be quite confusing for inexperienced users.

Such software can, however, give some indication of possible sources that can be investigated more thoroughly using other methods. Since, perhaps surprisingly, there will be different possible sources found by different systems, it can be advantageous to use more than one system. There are many factors that can influence the results, the most important being the choice of sample text or keywords to use for searching. In fact, the only time during the testing that exactly the same result was seen was when one system that was being evaluated was discovered to be sending the text to be examined on to *Turnitin* using a hijacked account. After replacing the *Turnitin* logo with their own logo and making some minor visual changes, the report was sent back to the user. This system is marketed to students, who pay a modest, one-time fee for having their paper checked against a system that their own university most probably will be using, as reported by Weber-Wulff & Köhler (2010b).

It is important to realize that some systems have trouble with large files. Some are not even able to read in the entire file; this can pose a problem when dealing with dissertations. For example, when testing the dissertation of Karl-Theodor zu Guttenberg with the *Turnitin* variant *iThenticate*, the thesis had to be split into 13 parts. Other systems, such as *PlagAware*, broke off after only about 160 pages (Weber-Wulff & Köhler 2011).

Another problem posed by large files is that only samples of the text are actually looked at, very seldom the entire document. The test by Weber-Wulff et al. (2013) showed that some systems capable of finding plagiarism in a three-page text do not find the identical plagiarism when it is embedded in 80 other pages of original (random) material. Very seldom are the systems able to make use of the material available at Google Books as potential sources.

There is also the problem of characters with umlauts, diacritical marks, or special character sets, as briefly mentioned above. Some systems will just ignore all words that have special characters in them. This gives dishonest students a simple method for foiling the systems: They only need to replace some individual characters such as 's' or 'e' with a character called a *homoglyph* from another alphabet. These are characters, often from the Cyrillic or Greek alphabets, that look the same or are very close in appearance, but which use different internal codes for their representation. A printed paper will look fine, as will the PDF. Copying text from the PDF will copy the internal representation, not the appearance, so not even Google can help here. Instead, one can look at the text in a word processor with the spelling check enabled. Since the words will not be in the dictionary, they will all be marked.

There are simple, free programs available for character substitution, so teachers need to be aware of this problem, although the makers of plagiarism detection software are working on solutions. During the 2013 test, *Turnitin* was seen as the only system that was able to properly deal with almost all of the homoglyphs, and

Urkund at least gave a warning that the text contained such characters. It is good that this is just a warning, as there can be many legitimate reasons for having such characters in a paper.

4.3.5 Documentation

A time-consuming part of a plagiarism accusation is the preparation of convincing documentation, as there will be numerous parties involved in possible sanctioning who will need to be convinced of the severity of the issue. A standard and easily readable form is to prepare a synoptic presentation, or side-by-side, giving the plagiarism on the left-hand side and a source on the right-hand side. It is imperative to record exactly the information necessary for a third party to verify that this is a true copy of each work. Bibliographic information (especially the edition used) must be recorded, and the page and line numbers of each of the excerpts should be documented in order to make it easy to quickly find the portions of the text parallel in the books.

This a major problem with so-called plagiarism detection software: They do not record any information about the page or line numbers, so one is compelled to search within the text in order to find the position of the fragment in question. One of the problems is that there is usually a difference between the physical page number of a PDF and the logical page of the document that is being investigated.

The documented portions should not be too fine-grained: A half-sentence here or there is probably not a plagiarism that will lead to rescinding a doctorate, although it might incur some sort of sanction in a term paper. But once the collection of fragments has grown to large proportions, then a full report needs to be prepared. It is important to keep in mind that numbers can be misleading. A full-page plagiarism should be more serious than a phrase here and there scattered over a number of pages. And the percentages given as scores by many plagiarism detection systems often do not state their basis. It is extremely important here to differentiate between percentage of the sample and percentage of the document. Experiments on Vroni-Plag Wiki have led to automatically generated overviews, but they also have many problems and are not always easy to interpret. There is currently no known tool for easily documenting and commenting plagiarism found.

4.3.6 Finding a Thesis

The discussion up until now has assumed that an educator has a copy of the term paper or thesis, at least in printed form, on his or her desk. Because of the wide press coverage of the plagiarisms found in the dissertations of some German politicians, the question has often arisen as to whether the dissertation of some particular person is plagiarized or not. This may be someone who particularly flaunts the use of the doctoral title, or is just an irritating person, or as in the case of the election of Pope Francis, someone who has recently attained a high office who is said to have obtained a doctorate in Germany (which he had not, but that is beside the point). No matter how morally problematic the reason for investigating a dissertation of someone may be, a dissertation (in Germany at least) is a public contribution to the body of academic knowledge and is available for scrutiny by anyone, not just members of a particular faculty of a particular university. As such, all dissertations defended in Germany must be published, so the first step of such an investigation is to consult the national library. For the German-speaking countries, these are the German National Library (n.d.), the Swiss National Library (n.d.), and the Austrian National Library (n.d.).

Unless the thesis was quite recent, the national library will have an entry for all dissertations that have been accepted in Germany since 1945, including the year of publication and the university at which the dissertation was defended. In addition, the national library keeps a copy of each thesis, either digitally or as a published volume that is in general not available for interlibrary loan. For older dissertations, there are yearly publications (Jahresverzeichnis n.d.) that list the dissertations and habilitations accepted – and sometimes list the dissertations rescinded for plagiarism or other reasons in the addenda in later years. Doctorates rescinded during the late 1930s and the early 1940s need to be carefully examined, however, as many were retracted on the basis of political reasons and not because of academic misconduct on the part of the author.

It is not necessary to know the title of the thesis or the university in question, as searches can be made by author name as well as by words in the title. The library catalog database entry will contain information about when and where the dissertation was accepted, as an example using the author's dissertation demonstrates in Fig. 4.2.

Link zu diesem Datensatz	http://d-nb.info/951730800
Titel	Contributions to mechanical proofs of correctness for compiler front ends / Debora Weber-Wulff. Institut für Informatik und Praktische Mathematik der Christian-Albrechts-Universität zu Kiel
Person(en)	Weber-Wulff, Debora
Verleger	Kiel : Inst. für Informatik und Praktische Mathematik
Erscheinungsjahr	1997
Umfang/Format	6, 186 S. : graph. Darst. ; 21 cm
Gesamttitel	Institut für Informatik und Praktische Mathematik <Kiel>: Bericht ; Nr. 9707
Hochschulschrift	Zugl.: Kiel, Univ., Diss., 1996
ISBN/Einband/Preis	kart.
Sprache(n)	Englisch (eng)
Sachgruppe(n)	28 Informatik, Datenverarbeitung

Screenshot: http://d-nb.info/951730800

Fig. 4.2 A dissertation listing from the German National Library

Germany has numerous online union catalogs, usually operated by the individual states. They can be used to see if there is a copy of the thesis in a library nearby. For example, in Berlin there is a union catalog for all the libraries in Berlin and Brandenburg, the Kooperativer Bibliotheksverbund Berlin-Brandenburg (n.d.) (KOBV), and the Karlsruhe Virtual Catalog (n.d.) (KVK), which allows a metasearch of many union catalogs, including foreign ones, to be undertaken. The Berlin State Library – Prussian Cultural Heritage (n.d.) also has extensive collections that are for the most part researchable online. There is also a worldwide union catalog that connects many library catalogs, *WorldCat*. Another possible source for a specific work is the online public access catalog (OPAC) of the university library where the dissertation was submitted.

Copies of most dissertations are kept in the institutional library or the university library of the university at which the thesis was accepted, and often copies can be found in other libraries. More recently, universities have taken to publishing dissertations online on their own open access servers.

Once a copy has been located, it can either be borrowed or, if it is not nearby, obtained using the interlibrary loan system. This is normally only possible for registered patrons of the library, but even university libraries will often permit local residents with valid identification to register. There is often a slight fee that is incurred for the interlibrary loan, but that is generally only to cover the postage. Germany also has services such as *SFX* or *Subito* that will attempt to determine if there is an obtainable digital copy of the thesis. *SFX* can be found integrated into the catalog entries in many of the university OPACs. An interlibrary loan may take a few weeks, but it is possible to obtain even rather old or obscure theses in this manner.

There are a number of aspects that need to be considered if the thesis cannot be readily found in Germany. For example, it is possible that the dissertation was submitted and granted under a birth name. Not only women change their name upon marriage, more and more men do the same. Sometimes people also change the use of their first names, either choosing a middle name or even eliding a complex name such as Friedrich Wilhelm into Friedhelm. And of course, if the person in question has a common name, there could be a dozen theses submitted by people of this name to various universities. It can be useful to find biographies of the person in question in order to get ideas as to where the thesis might be found.

Theses that were submitted to universities in other countries will not always be found so easily. In many countries there is no obligation to publish a dissertation, so it may be difficult or impossible to obtain copies of foreign ones. It used to be the case that foreign doctoral degrees needed to include the name of the university from which they were conferred, but as a result of the Bologna Process, degrees that were granted from an accredited university anywhere in the EU are able to be used just as "Dr." in Germany. The rules for the state of Berlin, for example, can be found in a document prepared by the Senatsverwaltung für Bildung, Jugend und Wissenschaft im Land Berlin (2013). In such a case, there may be no possibility for finding the written thesis, short of asking the person in question for a copy, but of course that cannot be easily done anonymously.

4.3.7 Mission Impossible?

It is quite a bit of work to find and document plagiarism, especially in a large text. There is no simple process to follow, but there are many heuristic methods that can be applied in the process of investigating whether a particular work is possibly taken from another source. Educators need experience in detecting plagiarism, so that they are more easily able to use the various tools and methods in order to avoid letting a plagiarism slip through.

There are some types of plagiarism such as translations or structural plagiarism that cannot be found by software. There are always lucky situations, of course. The grader might just happen to be reading the book a plagiarist was using, or a Google search only on proper names turns up the source in a different language. These types of plagiarism can best be found by a person reading the paper in question carefully with a solid knowledge of the published material available in this field. In the past, much plagiarism was discovered in just this manner – people reading a text and noticing how oddly similar it was to something they had perhaps just recently read themselves, or even finding their own words used by others. The case of Margarita Mathiopoulos, discussed in Sect. 3.5.8, is an example of one found by chance many years ago, although the university at that time dismissed the obvious text parallels that were documented.

Plagiarism detection software is only useful for catching the laziest plagiarists, those who copy large portions from online sources, especially from the Wikipedia. Even these are not always found by software and it is not always clear why such simplistic plagiarisms that tend to be found in student work, are not detected. The zu Guttenberg thesis, as described in Sect. 3.1, was one that used large chunks of text that were easy to find on the Internet. Just the same, software systems that were tested after the documentation had made clear the extent of the copying either gave the thesis a clean bill of health or turned up so many small bits that it was difficult to find the wheat for all of the chaff (Weber-Wulff & Köhler 2011).

Paraphrasing and extensive synonym or homoglyph substitution can make the search for a source quite difficult. Contrary to popular opinion, just changing a few words does not make the text original or usable without attribution. But this can have the effect of making it much harder to automatically detect the plagiarism. Luckily, people who plagiarize are often cutting corners in other places or are under time pressure and may at some point give up on rewriting, leaving a larger portion of the text unchanged. They also tend to not use spelling checkers and the like, so they will also copy errors from the original source into their plagiarized work. Looking for errors can lead to finding the source, despite rewriting efforts.

One example of using errors to find sources can be seen in the case Gc (Vroni-PlagWiki 2011b). This dissertation, submitted in the year 2000, wrote on page 51 about a provider called CompuServe, giving a book from the year 1997 as a source, footnoted at the end of the paragraph. In 1998, however, CompuServe had been sold to another company and had rapidly lost market share. Obtaining the book from 1997 showed that three entire paragraphs had been copied verbatim from this book, and indeed more than 70% of the pages turned out to be plagiarisms, much like Fig. 4.3.

Untersuchte Arbeit:
Seite: 051, Zeilen: 4-24.102

Damit das Internet nicht in Anarchie und Chaos versinkt, wurden etliche Kommissionen und Verwaltungseinrichtungen gegründet, die dringend notwendige administrative Aufgaben übernehmen. Zu den bekanntesten gehört das Network Information Center (InterNIC), das nationale Ableger hat, in Deutschland DE-NIC genannt. Das NIC registriert die im Internet offiziell gemeldeten Computer (Hosts) mit Namen und Rechnernummer. Hier werden zum Beispiel Namen wie „wdr.de" oder „aol.com" zentral registriert, damit sie garantiert nur einmal verwendet werden.

Eine weitere, offizielle „Behörde" ist die Internet Society (ISOC). Sie veranstaltet Kongresse, auf denen unter anderem kommende Standards diskutiert und schließlich auch verabschiedet werden. Eine spezielle Abteilung der ISOC, die sich Internet Architecture Board nennt, überwacht die technischen Aspekte des Netzes, veröffentlicht neue technische Spezifikationen (zum Beispiel Protokolle) und vergibt Rechneradressen. Alle anderen administrativen Aufgaben, die den Benutzer betreffen (wie oft wird das Internet genutzt, welche Adresse steht einem zur Verfügung, welche Nutzungsbeschränkungen gibt es etc.), übernehmen jene Dienstleister, die ihren Kunden den Zugang zum Internet verschaffen (Provider). Dies können kommerzielle Online-Dienste wie CompuServe, AOL oder T-Online sein, es kann sich aber auch um Unternehmen handeln, die „nur" Zugang zum Internet bieten. Wenn eine Firma Zugang zum Internet gewährt, werden die genannten Aspekte von den Administratoren des Unternehmensnetzes geregelt. 108

108 Schieb (1997), S. 15 f.

Quelle: Schieb 1997
Seite(n): 15-16, Zeilen: Abschnitt "Verwaltung des Internets"

Damit das Internet nicht in Anarchie und Chaos versinkt, wurden etliche Kommissionen und Verwaltungseinrichtungen gegründet, die dringend notwendige administrative Aufgaben übernehmen. Zu den bekanntesten gehört das Network Information Center (InterNIC), das nationale Ableger hat, in Deutschland DE-NIC genannt. Das NIC registriert die im Internet offiziell gemeldeten Computer (Hosts) mit Namen und Rechnernummer. Hier werden zum Beispiel Namen wie wdr.de oder aol.com zentral registriert, damit sie garantiert nur einmal verwendet werden.

Eine weitere, offizielle „Behörde" ist die Internet Society (ISOC). Sie veranstaltet Kongresse, auf denen unter anderem kommende Standards diskutiert und schließlich auch verabschiedet werden. Eine spezielle Abteilung der ISOC, die sich Internet Architecture Board nennt, überwacht die technischen Aspekte des Netzes, veröffentlicht neue technische Spezifikationen (zum Beispiel Protokolle) und vergibt Rechneradressen. Alle anderen administrativen Aufgaben, die den Benutzer betreffen (wie oft wird ins Internet gegangen, welche Adresse steht einem zur Verfügung, welche Nutzungsbeschränkungen gibt es etc.), übernehmen jene Dienstleister, die ihren Kunden den Zugang zum Internet verschaffen (Provider). Dies können kommerzielle Online-Dienste wie CompuServe, AOL oder T-Online sein, es kann sich aber auch um Unternehmen handeln, die „nur" Zugang zum Internet bieten. Wenn eine Firma Zugang zum Internet gewährt, werden die genannten Aspekte von den Administratoren des Unternehmensnetzes geregelt.

Fig. 4.3 A copy & paste plagiarism from the dissertation of *Gc* (VroniPlagWiki 2011c)

The example from Fig. 4.3 could have been called a pawn sacrifice (see Sect. 2.2.7) if the text had be paraphrased or reworded here and there. But the categorization is perhaps only of academic interest; what educators need to realize is that the source of the plagiarism is sometimes actually named in the thesis, either close to the copy via footnote or by mentioning a source only in the bibliography. Just checking a few footnotes – something that should be done anyway – can also result in promising sources that can be digitized and compared with the work at hand, if suspicions have been sufficiently roused.

Bibliometric analysis, for example, looking at the distribution of the references by year, or producing a word histogram in order to find misspelled words, or even calculating a character histogram so that the use of homoglyphs can be discovered, is not for everyone and most probably will not be one of the tools used in everyday plagiarism detection at university. But it can be used to produce some clues leading to possible sources.

Discovering plagiarism is not something that can be done with the push of a button. There is no quick, visual diagnosis possible, although once the plagiarism has been documented, it appears simple to see. However, as an educator gains experience with the task, good observation skills will raise a suspicion and it becomes easier and easier to use general research skills and simple software tools to find and document plagiarism, even in older, non-digital, or extensive works. The researchers at the VroniPlag Wiki have demonstrated this over and over again. The next section will give a deeper look into the work processes that take place in that context.

4.4 VroniPlag Wiki methodology

One often reads in the media that the VroniPlag Wiki researchers, introduced in Sect. 3.2, have some sort of advanced software that they use for rapidly finding plagiarisms in dissertations. On the contrary, the VroniPlag Wiki methodology involves much non-automated work and a multiplicity of methods, but in general follows the procedures laid out in Sect. 4.3 with a few added safeguards to ensure that the documentation, which can be prepared by anyone, is accurate and true. There are, however, a few software tools that can be used to help find and document text parallels that could be considered to be plagiarisms. This section will explain the major aspects of the workflow used by the group and how the tools work.

4.4.1 Technological Basis

A wiki is used as the major software basis for the platform. This is a system that permits geographically distributed users to be able to collaborate in a simple way. One of the most well-known uses of a wiki is the Wikipedia, which is often confused with the underlying technology. A wiki is just a collection of online pages that can easily be linked to each other. Each page in the wiki has an associated discussion page, and the pages can be grouped into so-called name spaces. Anyone, even someone who is not registered with the wiki, is able to edit the pages or contribute to a discussion or start a discussion in the accompanying forums area, if the wiki is set up to permit such actions. All versions of a page are preserved and can thus be viewed and referred to at a later point in time.

People can register a user name on a wiki. This will hide their IP address from general view and it serves to collect all of the edits that have been done under this name on this particular wiki. There is also a user page and a blog available to registered users so that they have a personal area for carrying out experiments or recording information that might not fit in the overall structure of the wiki. If a user does not register, then all of the edits are recorded with the IP address visible to anyone who wishes to look. This could reveal information about the person editing, for example, their university or place of employment.

The VroniPlag Wiki group uses a wiki run by the company *wikia.com* for organizing the documentation. Although the use of this wiki is without cost, this does mean that ads are run on the pages – and they unfortunately often enough end up being ads for ghostwriters or companies willing to help people obtain a doctorate if they pay a fee – but it has the advantage that someone else is responsible for the technical day-to-day operation of the wiki, as well as organizing backup and dealing with load imbalances and the like.

The technical platform for the wiki does have its problems. It is relatively simplistic and does not have many of the enhancements that have been developed for other implementations of the basic wiki software. Luckily, some of the VroniPlag Wiki users are computer science researchers and they have managed to build in the odd additional functionality that was deemed necessary for increasing the usefulness of the system.

Since it is still often useful to communicate with other collaborators in real-time, there is also an online chat available on the Internet Relay Chat (IRC) network *Freenode*. In the chat room #vroniplag, people gather or just hang out and haggle about how to interpret this or that text, and also post requests for others to help in obtaining literature. People who would like to learn more about plagiarism documentation also drop by with questions.

A number of other tools are also often used. These include scanners and OCR software, text extractors and comparing tools, automatic visualization and report generators. There are so-called bots that take care of tasks that can easily be automated, such as moving an entire documentation to a different part of the wiki. The openness of the wiki is very useful as it permits external programs to extract information that can then be used for various tasks.

One task that has to be done by hand is keeping a watch over the wiki. As it is an open platform, anyone is permitted to edit most of the wiki. Some persons think that they can remove information they do not like and vandalize some pages, emptying them of content or replacing them with more or less obscene or threatening material. Administrators of the wiki spend a good bit of time shutting out troublemakers, restoring pages, and removing material that has nothing to do with the intended purpose of the wiki. Trolls, as the vandals are often called on the Internet, appear not to understand that a wiki keeps copies of all versions of the wiki, so the work will not be lost if someone deletes a page. It can be restored from the backup with a single click.

4.4.2 Distributed, Collaborative Work

Since a wiki is a web-based system, the researchers do not need to be co-located in space or in time in order to communicate with each other. One of the advantages of a distributed research group working together is the wide range of materials that can be accessed. Since much research into the possible sources for a dissertation is conducted manually, it is necessary to obtain copies of material that can be rather obscure. VroniPlag Wiki contributors belong to different universities or other organizations and thus have access to a wide variety of online and offline materials.

People are sometimes even willing to drive to a neighboring town in order to obtain a reference from a library there, if it can not quickly be obtained by interlibrary loan.

The collaborative nature of the group is also beneficial for the effort because each of the members brings a different standpoint into the discussion and can contribute diverse talents to the tasks at hand. Active researchers come from many different fields and geographic locations. There are bookworms who appear to live in public libraries and weekend workers with access to digital ones; there are Google tamers who can dig out obscure footnotes and references and IT-savvy researchers who can force the technology to conform to the expectations of the group. Some enjoy looking for sources, others prefer documenting or proof-reading. It is this variance within the group of collaborators that gives it a strength far beyond what one individual could achieve alone.

4.4.3 Choice of Candidate Thesis

Although the impression is widespread that the VroniPlag Wiki platform mainly documents plagiarism in the theses of conservative politicians, it only seems that way because the press tends not to report about the text parallels in dissertations of non-celebrities. There is no targeting of politicians involved in the choice of thesis to investigate, but instead the group relies on input from the outside or chance discoveries. An anonymous drop box where people can send in information on suspicious, published university works is linked from the home page (Hushmail n.d.). However, the group will not act on statements such as "X is an idiot, can you have a look at their thesis?" A well-founded first suspicion will demonstrate text parallels; this will serve to interest others in looking for more and helping to documenting it. People who register a suspicion are also encouraged to participate in the documentation themselves. Quite a number of theses that have been found to have excessive text parallels have been made known to the group in this manner, and the informers have stayed on, documenting additional cases.

Other theses have been found by chance – while looking for some information, someone happens upon two sources for a particular statement that are worded identically, but written by different authors. The discoverers are often doctoral students who are making a thorough search of the literature and come up with a text they have already read somewhere else. It is understandable that they are often reluctant to inform the committee on good academic practice of their own schools, as they fear retribution. Sometimes a reader observes while studying a thesis that the text lurches violently in style from paragraph to paragraph, or that there is a particular statement or phrase that the reader recalls having seen before. With a little bit of research, another older text that is similarly (or identically) written can sometimes be found. In a worst-case scenario, readers finds paragraphs from one of their own publications under the name of someone else.

Documentation for at least three of the cases began because a researcher was in the process of writing a review of a publication, and happened upon text that the reviewer knew was from someone else, or which was so obviously jumpy in style, or which was so Wikipedia-like, that the reviewer became suspicious and began to check systematically.

4.4.4 Finding Sources

Sifting through Google results, as described in Sect. 4.2, is not the only way to come up with possible sources. The footnotes in the thesis can be closely examined. Is there a source that is excessively quoted? If so, this is a good source candidate. It can be obtained to see if perhaps more has been taken from this volume than what is indicated in the thesis. There seems to be a popular misconception, especially amongst students of law, that using "cf." (*vgl.* in German) means "This text was taken from that source, just reworded a bit" or that "cf. in particular" (*vgl. insbesondere*) means "This entire portion was taken word for word from this source."

Another good avenue for finding possible sources is to pretend to be researching the topic and to consult online databases. Many are searchable from anywhere in the web, and university libraries that subscribe to them will be able to provide digital copies of articles. For medical or biological topics, the *PubMed* database, run by the US National Library of Medicine, indexes the *MEDLINE* database as well as many journals and books in the field. Business or political science topics can be researched in Germany on the *Wiso-Net*, this database also includes many daily newspapers. Legal research can done via *Juris* or *Jurion*, and in mathematics, computer science or engineering there are digital libraries available from publishers such as *Springer*, *ACM*, or *IEEE* have digital libraries. *arXiv* lists many open access journals, and with the *Internet Archive* it is sometimes even possible to find web pages that have been removed or changed.

Some authors are also guilty of what could be called footnote plagiarism. That means that they have copied some or all of the references from a source. This source, however, is a secondary one, and the author is thus giving the impression of having worked intensively with the primary sources and read them, when in fact no effort has been made to obtain and read them. This can often be determined when the errors are copied as well, as noted above. Typical examples are statements that are not to be found at the given page, errors in bibliographic data, errors in translation or transcription, or even quoting sources that were never published. A flagrant bibliographic error can be quite useful as a search term, turning up additional probable sources.

Sometimes the important sources are the ones that are given in the text but not listed in the references. It is a good idea while reading a thesis to check a few of the footnotes given to see if they are listed properly in the works cited. A plagiarist will sometimes copy text along with the reference but forget to also include the paper referred to in the works cited. Searching for such orphaned references can sometimes give a clue as to a probable source because the source will have used this same reference. In addition, textbooks or other volumes that are not listed in the works cited can also be consulted. Some students assume that old texts are obscure and thus take text from them without realizing that they are standard material that may actually be correctly quoted in sources available online.

For some of the more difficult theses that were documented by the VroniPlag Wiki, the *brute force* method was used (see Sect. 4.3.1). All of the references listed in the bibliography were obtained, digitized, and compared with the thesis. This results in a good overview of the mosaic structure of the thesis and lends itself well to interesting visualizations. But using brute force does, however, involve scanning many thousands of pages and is thus extremely time-consuming, as well as demanding much storage space, so it is not feasible for general use, only for cases in which the documentation of every possible text parallel is deemed necessary.

4.4.5 Documenting Parallels

In documenting a text parallel, the VroniPlag Wiki group has found it best to use a basic unit called a fragment. This is a portion of the dissertation or other text under investigation that is maximally one page long and is from exactly one source. For each fragment, the portion of the text from the dissertation and the corresponding portion from the source are copied into a special page on the wiki. The page numbers and the line numbers for the beginning and end of each portion are also documented so that it is easy for someone to take the physical book and to quickly determine that a true copy has been made. An example of the documentation can be seen in Fig. 4.4, depicting the form used to generate the fragment shown in Fig. 4.1.

For some kinds of plagiarism, for example, shake & paste or mosaic plagiarism from multiple sources, this type of documentation is not practical, as one only sees tiny fragments that, taken on their own, may be considered too trivial to be considered plagiarism. Only in an overview would it be clear that this is regarded as plagiarism, but there is currently no good way to produce such an overview.

For each source determined, the VroniPlag Wiki contributors collect the bibliographic data on another special page. This page will also list all of the fragments that are found from this source. Fragments are then categorized into one of a few general categories: Copy & paste, disguised, pawn sacrifice, and sometimes even just no category, if the fragment is being documented for other reasons such as using unusual sources, for example, using a tabloid magazine as the source for population data, as is done in the thesis of *Jg*. The person setting up the fragment signs the fragment by adding his or her wiki name to the form, and can also add comments about the fragment in a free-text field. Before a fragment is published, that is, included in the documentation, it must go through a quality assurance process, called screening (*sichten*).

In the zu Guttenberg case on the GuttenPlag Wiki, the quality assurance process insisted on at least two persons verifying each fragment. In order to verify the fragment, a copy of the dissertation page and the source pages must be compared to the fragment documented. Often the second or third pair of eyes finds OCR errors

Fallkuerzel:	Dd
Typus:	Verschleierung ⬍
SeiteArbeit:	103
ZeileArbeit:	2-14
Quelle:	LAUBAG 1998
SeiteQuelle:	10
ZeileQuelle:	7-21
Markierungslaufweite:	4 ⬍
TextArbeit:	Der Tagebau Welzow-Süd ist der Hauptversorger des neu entstandenen Kraftwerkes Schwarze Pumpe mit einer Leistung von zweimal 800 Megawatt, das mit einem Wirkungsgrad von 40,5 Prozent dem neuesten Stand der Technik entspricht. Weiterhin liefert der Tagebau Welzow-Süd die Brikettierkohle für die Veredelungskapazitäten Schwarze Pumpe. Darüber hinaus bestehen weitere Aufgaben für den Direktabsatz, in der Zufuhr zu weiteren Kraftwerken der VEAG sowie in der Versorgung des Sekundärrohstoff-Verwertungszentrums Schwarze Pumpe. Um die erforderliche Planungssicherheit für einen längeren Zeitraum zu schaffen, hat die Landesregierung auf der Grundlage von Paragraph 12 Absatz 6 des Gesetzes zur Einführung der Regionalplanung und der Braunkohlen- und Sanierungsplanung im Land Brandenburg per Rechtsverordnung am 23. Dezember 1993 den Braunkohlenplan Welzow-Süd, räumlicher Teilabschnitt I, für verbindlich erklärt (Tagebau Welzow-Süd, Abbauentwicklung, siehe
TextQuelle:	Der Tagebau Welzow-Süd ist der Hauptversorger des neu entstehenden Kraftwerkes Schwarze Pumpe mit einer Leistung von 2 x 800 MW, das mit einem Wirkungsgrad von 40,5 Prozent dem neuesten Stand der Technik entspricht. Weiterhin liefert der Tagebau Welzow-Süd die Brikettierkohle für die Veredelungskapazitäten Schwarze Pumpe. Darüber hinaus bestehen weitere Aufgaben für den Direktabsatz, in der Zufuhr zu weiteren Kraftwerken der VEAG sowie in der Versorgung des Sekundärrohstoff-Verwertungszentrums Schwarze Pumpe. Um die erforderliche Planungssicherheit für einen längeren Zeitraum zu schaffen, hat die Landesregierung auf der Grundlage von § 12 Abs. 6 des Gesetzes zur Einführung der

Fig. 4.4 Documenting details of text parallels in VPW case *Dd*

the first person overlooked, or copies the emphasis more truly, or notices a stray hyphenation character that needs removing. Sometimes the verifier also sees how to extend the fragment.

In most of the cases on the VroniPlag Wiki there has been only one verifier, who also signs the fragment. This means that this person is also convinced that this piece of the documentation is indeed potentially a plagiarism. If it was the first fragment for a page, a special page is set up that will include all the fragments on this page. The verifier also checks to see if the page has now passed 50% or 75% of the lines and sets the category appropriately.

If a contributor has an issue with a fragment, a discussion is started on the discussion page in the wiki for that fragment. Some discussions are also held in the chat room, and there can be heated debates about whether a particular fragment is to be considered plagiarism or not or whether it is just too short to count. Sometimes being able to extend a fragment can lead to a resolution of such a dispute.

It is not ethical, in the eyes of the VroniPlag Wiki researchers, to name the author in full in public only on the basis of a few text parallels. Even if this might be enough to fail a candidate if discovered before a degree is awarded, a larger amount of text parallel is deemed necessary in order to rescind a doctorate, especially one that was granted years ago.

Since working with a public wiki means that each and every edit to the system is done in full view of anyone interested, the first documentation of a case is set up in a special section of the wiki in which every document carries a disclaimer that it is currently only under discussion, not documentation. All documents there only use an abbreviation for the name of the author. It is, of course, trivial to find out the name of the author by doing a bit of research. The main point is to keep Google from indexing the name and linking it to a possible accusation of plagiarism that has, as of yet, not been verified.

The form of documentation in a wiki is very good for automatically producing side-by-side documentation of colored fragments. However, if an educator wants to follow this process for documenting fragments of a thesis to be graded, he or she will not want to go to the trouble of setting up a public wiki, and unfortunately there is not yet software available for doing so locally. Using a word processor, a template can be set up that has two columns on a landscape page layout with some boxes at the top for recording page numbers and the name of the source. Each fragment that is found would be documented with a new page.

4.4.6 Public Polishing

On the VroniPlag Wiki, if it has been determined that there are enough text parallels for a case to be considered serious, the entire documentation is moved from the analysis portion of the wiki to a more public place in the main wiki. Moving the documentation is taken care of automatically by a small software system called a bot. There is no rule governing when a case is moved out into the public – and there is no rule governing when one is moved back into the analysis area (which is a rare occurrence that has, however, happened). Journalists have reported on this or that number of pages affected being a "rule," but it is more the question of how serious the text parallels are in multiple places than it is a particular absolute number of documented parallels.

The main idea is that there must more or less be a consensus that this documentation will eventually be made public. Often it can take weeks until most of the contributors agree to make one more public. The discussions are held both in a chat room (which is relatively private, although on occasion unknown persons have found their way in, copied discussions, and made them public) as well as in the forums area of the wiki.

The public polishing of a case will entail setting up numerous navigational aids – most of which can be done semi-automatically – and detailing the most serious aspects. This is important as a quick look at a random fragment might only yield a half a sentence that is copied and cause the reader to wonder why this is considered plagiarism. But this might be the start of a longish text parallel that continues on over the next pages. Also, interesting findings such as text taken without reference from the dissertation advisor or (in the case of a habilitation) from the dissertations of advisees, or bizarre mistakes in mathematical formulas or figures, are documented on such a page. They are useful for rapidly being able to find the interesting parts of a thesis, and they set up the basis for a final report, should one be prepared.

A barcode is used to visualize the extent of the text parallels. An example is given in Fig. 4.5, the legend is described in Table 4.1. The different levels of text parallel are drawn in different colors, and the barcode is produced directly from the documented and verified fragments in the wiki when a page is viewed with the bar code on it. It is important that this can be done automatically, as the number of fragments changes often. The current percentage of pages affected by plagiarism is also calculated and displayed.

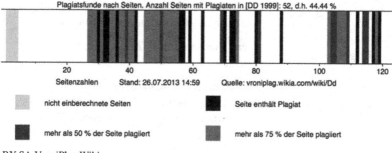

Fig. 4.5 The barcode for case *Dd* (VroniPlagWiki 2011d)

This barcode was originally intended as a help for the participants so that they could see which areas of the thesis have not yet been documented. But since journalists used the barcode excessively in reporting about the zu Guttenberg case, the bar codes have become iconic representatives for the plagiarism cases themselves, although they can suggestively skew the perceived amount of plagiarism to be more or less than is actually there. Many suggestions have been made as to how to improve them, but none seems to have caught on. Table 4.1 gives the meaning of the colors found in the barcodes now in use.

The wiki offers possibilities for preparing automatic reports and statistics that can be used to find additional errors in the documentation. Sometimes pages end up duplicated, and some pages are missing certain information or are not recording the page and line numbers properly. For a few of the cases, a formal final report was generated. For others, a simpler report listing all of the fragments and including the bar code and a textual description of the findings can be prepared within the wiki and stored as a PDF.

4.4.7 Publication

When the group of contributors feels that the case is serious enough to warrant publication, a bibliographic information page is set up with the author's name, the title of the dissertation, the name of the university and the department that granted the title, the names of the members of the committee, the date of the examination,

Table 4.1 Color legend for VroniPlag Wiki barcodes

Stripe color	Meaning
Black stripes	mean that text parallels have been found on the page.
Dark red stripes	mean that more than 50% of the lines on the page are documented in fragments.
Bright red stripes	mean that more than 75% of the lines are affected.
Blue stripes	denote the areas that are not part of the investigation, such as the table of contents, an appendix, empty pages in the middle of the thesis, etc.
White stripes	do not mean that this page is free of plagiarism. It may indicate that this page has not yet been looked at, or it may indicate that nothing has been found, although attempts have been made.

a link to the catalog entry in the German National Library, and a link to an online version, if available.

The full name and other statistical data is now entered into a table[6] and an announcement of a new case is made on Twitter. This does not mean that the case is now completed, it means that it is serious enough to be made public. Oftentimes the work will continue, eventually doubling or tripling the amount of pages found to contain text parallels.

VroniPlag Wiki does not as a body officially inform the universities, as there is no spokesperson for the group. However, individual contributors will often take it upon themselves to write to the university in question, either to the ombud for good academic practice, if that person can be determined from the home page of the university, or the president or rector of the university, or the dean of the department, or all three at the same time. Some also inform the accused plagiarist as a professional courtesy and journalists who are interested in plagiarism cases. Media interest – which is unfortunately only really awakened for cases that involve politi-

---

[6] There is both a general overview page (VroniPlagWiki 2013a) that includes a sortable table with information on the field, university, dates of publication and so on, as well as a visual timeline. A statistics page (VroniPlagWiki 2013b) includes many additional details such as the number of pages, footnotes, sources found, fragments, links to biographical information, and other information as can be determined.

cians – serves as a catalyst to get the investigation at the university going quickly. If this media attention is lacking, cases can and do drag on for years. It is not always the case that the individual contributor who informs the university is informed of the decision of the university, despite there being passages to this effect in the procedural rules of those universities.

Quite a number of persons who have had their doctorate rescinded have taken the universities to court. To date, the universities have won almost every case. It is quite interesting to read the legal judgments, as they are published and contain many salient details about the person in question, even if an attempt is made at anonymization. Roland Schimmel discusses some of these cases at length in an online law journal (Schimmel 2013). It is interesting to see that it has been specifically determined by a court in Cologne that a university is perfectly within its rights to use the VroniPlag Wiki documentation for investigating a case (Verwaltungsgericht Köln 2012a). However, not all universities have availed themselves of this documentation. Some universities have failed to rescind doctorates, as discussed in Sect. 3.2, although the extent of the text parallels is quite alarming. It is unfortunate that these universities have chosen to just state that they do not feel that the amount of text parallel warrants a rescinding of the doctorate. It would contribute more to the academic discussion if, at least for a few example fragments, the university committee could explain why this type of misattribution was not found to be problematic. That would be quite useful in furthering the discussion about what constitutes plagiarism.

If a doctorate is revoked by the granting university, the title will be struck from the documentation. At some point the case will be removed from the main page, but it is still available online and linked to from various overview pages. Fig. 4.6 depicts the status of the cases as of 26 July 2013. At that time there were 43 dissertations, three habilitations, one master's thesis, and one book on academic writing documented. Fourteen degrees have been revoked to date, but some are still being contested in court. Four court cases have been decided in favor of the university. In seven cases the universities decided not to revoke the thesis, but did not explain why the text parallels found are acceptable. The book on academic writing has been withdrawn by the publisher.

4.4.8 Problems with VroniPlag Wiki

The academic community in Germany, and especially the university system, has not welcomed the public documentation done at the VroniPlag Wiki. There has also been much misunderstanding, often spread by the media, about what exactly the group does, who its members are, why they work, etc.

The major problem with this collaborative group is that since it is not organized, there is no spokesperson. The media are used to having someone they can pepper with questions, so when someone who is not affiliated with the group begins pretending to be the speaker for the group, or when one active member answers questions as an individual, this will often be interpreted as the opinion of the group.

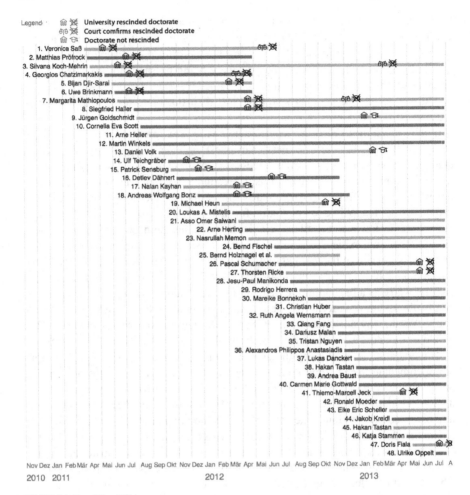

Fig. 4.6 Timeline of VroniPlag Wiki cases

People doing technical jobs such as being an administrator, which only entails having access to more advanced technical tools of the wiki system, are often thought of as being somehow in charge of the system. The new, emerging, non-hierarchical, collaborative work efforts that the Internet enables are not widely understood. Similarly, web-based projects with radically different goals that are also concerned with plagiarism are also often thought to be the same group. The distinction between different groups on the Internet is not easy for many to see.

The process of reporting to universities has turned out to be very problematic. The German ombud system was set up for whistleblowers, people with non-public information about academic misconduct. The ombuds are more often than not charged with mediating conflicts, not with independently investigating cases. The universities themselves often do not have detailed procedures for dealing with misconduct cases, especially not those that can be documented using publicly available material. Often the persons at the universities who are informed about a case have only understood that there is an allegation of misconduct. They do not follow the links given and look at the documentation, because they do not understand the nature of the wiki. VroniPlag Wiki has been very bad at communicating this structure to the academic community. It seems that investing time in preparing linear reports that combine all the information documented is useful, as a document with over 100 pages of colored text parallel is more believable than one link to somewhere on the Internet.

Of course, the name of the group is also unfortunate, as it is based on the nickname of the first person with documented plagiarism in her dissertation, although the focus of the group is solidly on the texts themselves and not the persons. No one could have foreseen in 2011 that there would be so many dissertations with plagiarism that were so easy to document. However, a name that is so widely, even internationally, known and recognized is not easily changed.

The academic establishment has responded to the press attention to prominent plagiarism cases by attempting to infer that publicly discussing academic misconduct is in and of itself academic misconduct, if the charges are not made in good faith. Some universities play down cases that have been made public, and cloak their proceedings in secrecy, but experience has shown that there is a danger, when such cases are not transparent, that they can drag on for a long period of time. Nonetheless, the responsible conduct of research demands that investigations be undertaken promptly, the results published, and attempts be taken to identify and repair the problem that led to the misconduct. There are no universities without academic misconduct. Those that say they do not have cases are not looking and listening closely enough. Excellent universities understand this, and see the investigation of cases of academic misconduct and plagiarism as part of their quality assurance program. They communicate their findings transparently as part of their striving toward constant improvement.

4.4.9 How has IT Changed Plagiarism Detection?

One of the most dramatic results of the work of the VroniPlag Wiki group has been the combination of available technology in order to permit plagiarism detection and documentation that has not been previously possible without an enormous investment of time and money. Much of the technology has been available for years or even decades, but the sum of using it in an interconnected manner is much larger than the parts.

The Internet has been available for over 25 years, offering simple real-time communication via chat systems. Search engines have been indexing online material for almost as long as the Internet has existed, and with Google's large-scale scanning of printed books, the tools for quickly finding resources are available, if people learn how to formulate good search queries.

With the availability of cheap hard disks for dealing with large files, good scanning and intelligent OCR technology, and a simple-to-use Wiki as a basis for collaborative editing, the technological basis for a group of people working together in the area of plagiarism detection is given.

Before the advent of the World Wide Web, an examiner would need to be lucky or have an excellent memory in order to find plagiarism in a thesis, although there must often have been suspicions based on shifts in writing style that just could not be proven. But now, even working alone, IT-based plagiarism investigations can be conducted without an enormous investment of money.

4.5 Collusion Detection

There is a specific kind of plagiarism that is said to be often found at universities, although as for all kinds of student misconduct there are no hard figures. This is the situation when two or more students submit either the same paper, or very similar papers that have only had a number of superficial modifications done to them, as the partial fulfillment of the requirements for a grade or degree.

This type of behavior is called *collusion*, in the sense of working together in a situation in which individual work is expected. In August 2012, for example, Harvard University announced that it was investigating a large group of students who had colluded on a take-home exam, although the instructions had explicitly stated that they could use the Internet, but had to work alone on the solutions (Pérez-Peña & Bidgood 2012; Robbins 2012).

It can be difficult to draw the line between permitted collaboration and not permitted collusion, as Barrett & Cox (2005) have noted. According to Barrett & Cox, instructional settings that have a high student-to-teacher ratio are more prone to have collusion happen. It can be intentional on the part of the student, for example, when the students assume that the busy teachers will not really be reading the papers or will not notice that the same or a similar paper has also been submitted. It can happen when students are not aware of the rules in place in the school about the copying of text, especially if the text was co-generated. It is also a large problem in computer science courses, where students solve small problems in a programming language. They are encouraged to work together in order to find a solution, but are expected to submit their own work. Often students will send in identical or quite similar code, as they are often convinced as students that there is only one way to solve a problem and do not realize that there are many ways to program a solution to a problem.

In 2012, researchers at the HTW Berlin tested a special case of plagiarism detection systems, so-called collusion detection systems (Weber-Wulff, Köhler, & Möller 2012). Software that can find collusion faces a different task than a plagiarism detection system, although some systems can be used for both tasks. While plagiarism detection is looking for possible sources in a database or on the open Internet, a collusion detection system works with a closed set of files. They are compared against each other to find similarities between the works, not to find possible sources. This kind of plagiarism is easier to find because there is just a small, finite set of documents to examine.

In order to test these systems, two sets of test cases were prepared. One set contained short essays that are in some way variations of an original essay. Some had original portions inserted at different points in the essay, others had words substituted with synonyms. There were various degrees of substitutions used, and a few essays were identical copies with homoglyphs replacing certain characters.

The other set was a collection of computer programs based on a collection of anonymized true student solutions to a programming exercise. In addition to 21 student solutions (a few had to be excluded as they were determined to have been the results of collusion that went unnoticed, although the students were told beforehand that their results would be used for a collusion test set), 22 additional solutions were prepared. Each additional solution used one technique for modifying the code, for example, renaming variables or methods or moving code around or inserting comments. The goal was to see which systems were able to find which kinds of collusion.

In assessing the field of candidate systems for the test, 33 products were found. However, 15 of these could not be used or the companies did not respond to emails requesting access, so only 18 systems could be tested.

One of the major problems that a collusion detection system must deal with is the explosion of the number of cases that need to be tested: each paper must be compared with all of the others in the group. That means that the number of tests that have to be done will increase rapidly as the number of papers being tested goes up. Additionally, the reports must be visualized so that candidates for collusion are immediately identifiable. Especially in a case where there is a collusion group involving more than two papers, it becomes quite difficult to represent the situation.

The identification of collusion in text ought to be trivial for a good plagiarism detection system. However, there are numerous small problems that can result in false negatives. That means that the system misses some obvious collusion. For example, some systems do not deal well with non-ASCII characters such as umlauts or other characters with diacritical marks and are easily confused when double blanks are inserted. Other systems have problems with text that has been edited so that they are not identical copies of each other. And of course, translations and strongly edited texts, although sometimes easily identifiable by a teacher, are quite difficult for a computer to identify.

Only four systems reached 20 points or more out of 28 on the text collusion test, *Turnitin*, *Ephorus*, *SPLaT*, and *JPlag*. The results for the program code collusion test were quite different from the pure text test. Due to the number of tests conducted – there were 22 test cases – 63 points were possible. However, only four systems earned more than 40 points on this part of the test: *JPlag*, *MOSS*, *Sim_Text*, and *SPLaT*.

The details of the test can be found online (Weber-Wulff et al. 2012).

References

All Answers Ltd. (2012). *How does Viper use my essay/dissertation?* [Web page]. Available at http://www.scanmyessay.com/viper-use-essay.php cited 21 December 2013.

Austrian National Library /ÖNB. (n.d.). *QuickSearch: die Suchmaschine der Österreichischen Nationalbibliothek.* [Web page]. http://search.obvsg.at/ONB cited 22 August 2013.

A.V. et al. ex. rel. Vanderhye v. iParadigms, LLC, 562 F.3d 630 (4th Cir. 2009). [Legal ruling], 16 April. Available at http://caselaw.findlaw.com/us-4th-circuit/1248473.html cited 24 July 2013.

Barrett, R. & Cox, A. L. (2005). 'At least they are learning something': the hazy line between collaboration and collusion. In: *Assessment and Evaluation in Higher Education*, Vol. 30, No. 2, pp. 107–122. Available at http://www.tandfonline.com/doi/full/10.1080/0260293042000264226 cited 3 March 2013.

Berlin State Library – Prussian Cultural Heritage. (n.d.). *The Online Catalogue of the Berlin State Library – Prussian Cultural Heritage – StaBiKat.* [Web page]. http://stabikat.de/DB=1/LNG=DU/SID=001cb108-2/LNG=EN/ cited 22 August 2013.

DIY Book Scanner. (n.d.). *Welcome to DIY Book Scanner.* [Web site]. http://www.diybookscanner.org/ cited 22 August 2013.

German National Library/DNB. (n.d.). *Catalogue of the German National Library.* [Web site]. http://www.dnb.de/EN/Kataloge/kataloge_node.html cited 22 August 2013.

Google. (2013). *Programmer's Guide > JSON/Atom Custom Search API > Overview > Pricing.* [Web page]. https://developers.google.com/custom-search/v1/overview cited 16 August 2013.

Grune, D. & Huntjens, M. (1989). Het detecteren van kopieën bij informaticapractica, In: *Informatie*, Vol. 31, No. 11, pp. 864–867. English translation available at http://dickgrune.com/Programs/similarity_tester/Paper.ps and the program code at http://dickgrune.com/Programs/similarity_tester/ cited 3 March 2013.

Hage, J., Rademaker, P., & van Vugt, N. (2011). Plagiarism detection for Java: a tool comparison. In: G. van der Veer, P. Sloep, & M. van Eekelen (Eds.) *Computer Science Education Research Conference* (CSERC 11). Open Universiteit, Heerlen, The Netherlands, 7–8 April, pp. 33–46.

Halstead, M. H. (1977). *Elements of Software Science.* New York: Elsevier.

Howard, J. (2012). Google Begins to Scale Back Its Scanning of Books From University Libraries. In: *The Chronicle of Higher Education (Online)*, 9 March. https://chronicle.com/article/Google-Begins-to-Scale-Back/131109/ cited 24 July 2013.

Hushmail. (n.d.). *Secure contact form (vroniplag).* [Web page]. https://forms.hush.com/vroniplag cited 22 August 2013.

iParadigms. (2011). *User Agreement.* [Archived web page], 13 April. http://web.archive.org/web/20110413080743/http://turnitin.com/agreement.asp? cited 22 August 2013.

iParadigms. (2013a). *User Agreement*. [Web page]. https://turnitin.com/agreement. asp cited 22 August 2013.

iParadigms. (2013b). *About us – our company*. [Web page]. http://www.turnitin. com/en_us/about-us/our-company cited 24 July 2013.

IPR Helpdesk. (2004). *Darf ich Aufsätze von Studierenden einfach an einen Plagiat-stestdienst weiterschicken oder brauche ich das Einverständnis der Studierenden?* [Web page]. http://plagiat.htw-berlin.de/ff/support/5_2/einreichungsdienste cited 24 July 2013.

Jahresverzeichnis. (n.d.). The series has changed names often:

1887–1913: *Jahresverzeichnis der an den deutschen Universitäten erschienenen Schriften*;

1914–1925: *Jahresverzeichnis der an den deutschen Universitäten und Techni-schen Hochschulen erschienenen Schriften*;

1926–1936: *Jahres-Verzeichnis der an den deutschen Universitäten und Hoch-schulen erschienenen Schriften*;

1936–1978: *Jahresverzeichnis der deutschen Hochschulschriften*;

1978–1987: *Jahresverzeichnis der Hochschulschriften der DDR, der BRD und Westberlins*

Karlsruhe Virtual Catalog. (n.d.). *KVK - Karlsruhe Virtual Catalog*. [Web page]. http://www.ubka.uni-karlsruhe.de/kvk_en.html cited 22 August 2013.

Kooperativer Bibliotheksverbund Berlin-Brandenburg. (n.d.). *KOBV-Portal*. [Web page]. http://digibib.kobv.de/ cited 22 August 2013.

Lancaster, T. (2003). *Effective and Efficient Plagiarism Detection*. [PhD thesis], London South Bank University, London, UK. Available at http://bcu.academia.edu/ThomasLancaster/Papers/79311/Effective_and_Efficient_Plagiarism_Detection cited 12 March 2013.

Lancaster, T. & Culwin, F. (2005). Classifications of plagiarism detection engines. In: *ITALICS*, Vol. 4, No. 2. Available at http://www.ics.heacademy.ac.uk/italics/download.php?file=italics/Vol4-2/Plagiarism%20-%20revised%20paper.pdf cited 12 March 2013.

Lukashenko, R., Graudina, V., & Grundspenkis, J. (2007). Computer-based plagia-rism detection methods and tools: an overview. In: B. Rachev, A. Smrikarov & D. Dimov (Eds.) *Proceedings of the 2007 International Conference on Com-puter Systems and Technologies* (CompSysTech '07), 14–15 June. Available at https://dl.acm.org/citation.cfm?doid=1330598.1330642 cited 7 July 2013.

Mozgovoy, M., Kakkonen, T., & Cosma, G. (2010). Automatic Student Plagiarism Detection: Future Perspectives. In: *J. Educational Computing Research*, Vol. 43, No. 4, pp. 511–531.

Peng, Z. (2013). *PRISM break*. [Web page]. *https://prism-break.org/* cited 22 Au-gust 2013.

Pérez-Peña, R. & Bidgood, J. (2012). Harvard Says 125 Students May Have Cheated on a Final Exam. In: *The New York Times*, 30 August. Available at http://www.nytimes.com/2012/08/31/education/harvard-says-125-students-may-have-cheated-on-exam.html cited 19 January 2013.

Robbins, R. (2012). Harvard Investigates "Unprecedented" Academic Dishonesty Case. In: *The Harvard Crimson*, 30 August. Available at http://www.thecrimson. com/article/2012/8/30/academic-dishonesty-ad-board/ cited 19 January 2013.

Schimmel, R. (2013). Zwei Jahre Plagiatsaffären: O-Töne und Originelles vor Gericht. In: *Legal Tribune Online*, 16 February. http://www.lto.de/recht/ hintergruende/h/plagiate-schavan-guttenberg-doktortitel-entzogen-klagen/ cited 22 August 2013.

Senatsverwaltung für Bildung, Jugend und Wissenschaft im Land Berlin. (2013). *Führung ausländischer Hochschulgrade, -tätigkeitsbezeichnungen und -titel.* http://www.berlin.de/imperia/md/content/sen-wissenschaft/berliner_ hochschulen/fuehrung_von_auslaendischen_hochschulgraden.pdf cited 2 July 2013.

Swiss National Library. (n.d.). *Quick Search to Helveticat.* [Web page]. http://opac. admin.ch/cgi-bin/gw/chameleon?skin=helveticat&lng=en cited 22 August 2013.

Vanderhye *ex rel.* A.V. v. iParadigms, LLC, No. 07-0293 (E.D. Va., March 11, 2008). [Legal ruling]. Available at http://www.steptoe.com/attachment.html/3370/501b. pdf cited 24 July 2013.

Verwaltungsgericht Köln. (2012). *Urteil VG Köln 6 K 6097/11.* [Legal ruling] 22 March. Available at http://www.justiz.nrw.de/nrwe/ovgs/vg_koeln/j2012/6_ K_6097_11urteil20120322.html cited 2 July 2013.

VroniPlag Wiki. (n.d.). *Quelle:Textvergleich.* [Web page]. http://de.vroniplag.wikia. com/wiki/Quelle:Textvergleich cited 22 August 2013.

VroniPlag Wiki. (2011a). *Dd > 103 > Fragment 103 02.* [Web page]. http://de. vroniplag.wikia.com/wiki/Dd/Fragment_103_02 cited 22 August 2013.

VroniPlag Wiki. (2011b). *Eine kritische Auseinandersetzung mit der Dissertation von Georgios Chatzimarkakis: Informationeller Globalismus. Kooperationsmodell globaler Ordnungspolitik am Beispiel des Elektronischen Geschäftsverkehrs.* [Web page]. http://de.vroniplag.wikia.com/wiki/Gc cited 22 August 2013.

VroniPlag Wiki. (2011c). *Gc > 051.* [Web page]. http://de.vroniplag.wikia.com/ wiki/Gc/051 cited 22 August 2013.

VroniPlag Wiki. (2011d). *Eine kritische Auseinandersetzung mit der Dissertation von Prof. Dr. Detlev Dähnert: Bewältigung technischer und sozialer Probleme bei der Konzeption von Umsiedlungen.* [Web page]. http://de.vroniplag.wikia.com/ wiki/Dd cited 22 August 2013.

VroniPlag Wiki. (2013a). *Übersicht.* [Web page]. http://de.vroniplag.wikia.com/ wiki/Übersicht cited 22 August 2013.

VroniPlag Wiki. (2013b). *Statistik.* [Web page]. http://de.vroniplag.wikia.com/wiki/ VroniPlag_Wiki:Statistik cited 22 August 2013.

Weber-Wulff, D. (n.d.). *Test of Plagiarism Software.* [Web site], prepared with assistance from Wohnsdorf, G., Pomerenke, M., Köhler, K., Möller, C., Touras, J., Zarzecki, M., & Zincke, E. http://plagiat.htw-berlin.de/software-en cited 22 August 2013.

Weber-Wulff, D. (2007). *Fremde Federn Finden.* [Web site], prepared with assistance from Wohnsdorf, G., Pomerenke, M., Köhler, K., Möller, C., & Zincke, E. http://plagiat.htw-berlin.de/ff cited 15 August 2013.

Weber-Wulff, D. & Köhler, K. (2010a). *Test 2010: S10-03 PlagiarismFinder*. [Web page]. http://plagiat.htw-berlin.de/software-en/2010-2/s10-03-plagiarismfinder/ cited 23 July 2013.

Weber-Wulff, D. & Köhler, K. (2010b). *Test 2010: S10-20 iPlagiarismCheck*. [Web page]. http://plagiat.htw-berlin.de/software-en/2010-2/s10-20-iplagiarismcheck/ cited 23 July 2013.

Weber-Wulff, D. & Köhler, K. (2011). Kopienjäger: Cloud-Software vs. menschliche Crowd in der Plagiaterkennung. In: *iX: Magazin für professionelle Informationstechnik*, No. 6, pp. 78–82. Available at http://www.heise.de/ix/artikel/Kopienjaeger-1245288.html cited 24 July 2013.

Weber-Wulff, D., Köhler, K., & Möller, C. (2012). *Collusion Detection System Test Report 2012*. [Web site]. http://plagiat.htw-berlin.de/collusion-test-2012/ cited 2 July 2013.

Weber-Wulff, D., Möller, C., Touras, I., & Zincke, E. (2013). *Plagiarism Detection Software Test 2013*. [Web page]. http://plagiat.htw-berlin.de/software-en/test2013/report-2013/ cited 23 December 2013.

Chapter 5
Plagiarism Perspective

Universities have always had to deal with problems of plagiarism. Some are open and transparent about what they do, others treat it as a shameful or secret thing that needs to be quickly swept under the carpet. Since educators want to focus on teaching their subjects and not on detecting plagiarism, there is a strong desire for a simple, computer-based solution. However, plagiarism is a social problem that is not easily solved by using software.

It cannot be repeated often enough that it is simply not possible to certify a paper or a thesis as being **free** of plagiarism. There is always the possibility that the source has just not yet been found – for example, because it is from an expensive expertise that is not held in a library or because it is from a book that is not listed in the bibliography. Some plagiarism detection systems announce "originality scores" or they present their customers with a "100% plagiarism free" certificate. As with errors in programs, it is only possible to demonstrate the presence of text parallels that might be considered plagiarism, not the absence.

> There is no method for proving the **absence** of plagiarism.

So since it is difficult or even impossible to detect plagiarism after it has happened, it is important that schools and universities increase their efforts to avoid plagiarism happening in the first place.

This chapter discusses various aspects of a perspective on plagiarism that deal effectively with some aspects of the problem. There are three major areas that need to be addressed:

- *Training* of students in the art of academic writing and proper quoting and the training of teachers in careful reading and discovery tactics for finding plagiarism
- Setting up *transparent processes* for dealing with plagiarism and academic misconduct as well as enabling an efficient investigation into suspected cases of plagiarism

- Giving teachers a service staffed by trained people that can be used for *testing* suspicious papers and theses

It is vital that the university administrations are dedicated to dealing with this problem. There must be a top-down commitment to good academic practice and not just lip service given, and there must be sufficient resources available for making it happen.

5.1 Training

As can be seen from the investigation of Franklyn-Stokes and Newstead (1995), a fear of punishment does not deter students from cheating. So investing much effort in finding and punishing students for cheating may not have as much payoff as the following methods. Training is indispensable, not just for graduate or undergraduate students, but for teachers and administrators as well. This section outlines some of the training that is necessary.

5.1.1 Introductory Lesson

Appendix D describes one idea for an introductory lesson and is based on a course first proposed in (Wilson & Ippolito 2007). This 90-minute introduction to the topic of plagiarism is ideal for use during an orientation week at the beginning of the first semester. It is not just a lecture, but explores together with the students a definition of plagiarism, exposes them to citation methods used in their field of study, introduces good working habits, and offers the students a chance to ask questions that may be bothering them. Students can, however, be very reluctant to ask questions in person, so passing out notecards for anonymous question-gathering can help overcome this problem. It can be useful to continue with discussions about their expectations of what university will be offering them and what will be expected from them, and perhaps also going into detail on tools for developing good study habits.

The art of note-taking has become quite important to be taught to beginning students as many will tend to multi-task in the classroom, dividing their attention between multiple running applications on their laptops/telephones and the teacher. Even those who are attempting to take notes on paper will have the problem of the people sitting in front of them watching amusing trailers or looking at pictures or engaging in other activities with their laptops. This can be quite distracting for those trying to concentrate, but is generally not realized to be a problem by the people sitting in the front.

A recent study (Risko, Buchanan, Medimorec, & Kingstone 2013) looked closely at the problem of mind-wandering during lectures and how it is enhanced by the tempting presence of an online laptop. They designed studies to investigate the question and conclude (2013, p. 280):

> Overall, the present results are consistent with the idea that trying to listen to a lecture while engaging in computer mediated non-lecture related activities will impair lecture retention.

However, keeping students from using their laptops can also be problematic, as some students will state, research results notwithstanding, that they find that it is easier for them to take notes using a computer.

If the students can learn how to take notes instead of just copying material, i.e. trying to scribble down exactly what is written by the teacher on the board or presented as slides, then they learn a valuable lesson. When they attempt to make their own meaning from the original – while still retaining a reference to the original – then they have taken the first step into learning how to incorporate someone else's work into their own.

It may sound strange to be dealing with apparently off-topic aspects such as how to take notes or how to organize one's studies, but it is more and more the case that students who are unprepared for studying are being admitted to university. Instead of complaining about the circumstances, it is best to confront the situation head on and offer at least introductory lessons, as well as communicating the school policy on plagiarism during orientation week.

If there is more time available, then it would be useful to speak with the students about the nature of the academic process and to discuss with examples why proper attribution and full transparency of sources is not just a formality, but the heart of science. Academic integrity forms the basis for all academic endeavors. Some illustrative examples of continuing replication of mistakes without correct attribution can be used, for example the Mountains of Kong in Africa (Bassett & Porter 1991, p. 367) or the maps depicting Baja California as an island (Library of Congress 2010).

5.1.2 Propädeutikum

In the humanities, there is a long tradition of incorporating a mandatory introductory course about the field of study into the canon. In this course, the students are introduced to the university library, to research tools and methods of their chosen field, and they are often given the opportunity to write example papers that are given plentiful feedback. Such an introductory seminar is often called a *Proseminar* or a *Propädeutikum* in Germany.

Many other fields such as engineering or medicine often do not have any sort of mandatory course teaching the basics of academic investigation at the beginning of the study program – rather, there is often a seminar on academic practice given in parallel with writing the final thesis. This is far too late, however, to be providing this information. Universities need to start offering such courses in the first semester, although this is quite a problem in the typical three-year bachelor's degree programs that are found in Germany, for example. Departments already complain that there is not enough time in just three years to fit in all of the introductory work, the social skills classes and internships now being introduced, and then to have time for the actual classes about the subject matter.

Perhaps offering beginning students an introductory semester or even an entire year in which they take composition, foreign languages, maths, and some basic courses in various subject fields that could then be credited towards the program they eventually join would be quite useful for helping students orient themselves and to learn the basic means of conducting and documenting academic inquiry. This is especially necessary as most German states have dropped back from 13 to 12 years of instruction in high schools. Of course, it would be necessary for the students to be able to have student status and obtain student loans for this additional year for such a scheme to be workable, and the universities would need increased resources to incorporate the extra students.

5.1.3 Every Class, Every Semester

Good academic practice is so important that the message about how to do it right bears repeating. The students should be reminded of what is expected of them in every class and in every semester. Each class needs to clearly delineate what is expected from the students in the ways of acceptable collaboration and what will be considered unacceptable. This goes beyond the general rules that are traditionally listed in a module description or are part of the university-wide examination rules.

For example, an introductory course in programming might require that students discuss the problems with each other, but that each student turns in his or her own copy of the results, representing work that they did by themselves. In an advanced course, it might be permissible for groups of two or three to hand in just one copy of the results for the group, but the students might be requested to clearly state which of them did which part of the work. Students may also be permitted to submit code from other students – as long as they give credit to the student they obtained the code from, and explain what they did themselves to adapt or build on this code. This is a good exercise in giving credit.

The reminder of what good academic practice is, about the expectations in the context of this particular course, and what constitutes acceptable collaboration needs to be well-documented in a syllabus that is either given to the students in writing or made available digitally at the beginning of each course. It would be advantageous for each university to provide their own materials, for example, a web site, a leaflet, or a handbook detailing the academic process, the need for good academic practice, and the procedures and consequences in place at that particular institution. Instead of hoping that the students will somehow know how to behave correctly, this would demonstrate that the university cares enough about good academic practice that they have developed their own material. It should go without saying that this material may not in any way be a plagiarism of materials used at other schools, although examples of plagiarism in a document about plagiarism have been found.

5.1.4 Design Plagiarism Out

Assigning the same laboratory exercises that will determine identical results or assigning the same topics for term papers year after year will increase the temptation to obtain copies of successful materials from a previous year and submit them. Teachers should sit down together with their colleagues and come up with new ways of assessing the learning that is going on so that plagiarism is designed out of the exercises and exams that are to be administered. If there is no possible source to plagiarize for a task solution, the only way to solve it will be to actually do it.

One method is to focus more on process than on product. Jude Carroll speaks of "less find and fake and more do and make" (J. Carroll, personal communication, 30 January 2013). This can have the added attraction of making students write reports in complete sentences, which trains important skills that often receive too little attention among science and engineering students. The students can be required to describe the goal of the exercise and the steps that they took in order to achieve that goals, instead of just handing in the final results. The data that was gathered or the programs that were written can be included in an appendix.

This, of course, puts a higher burden on teachers, as they have to read and comment on the reports submitted. The author has been doing this for programming exercises for over a decade and has had good results with it despite a bit of reluctance on behalf of some of the students who protest about having to write complete sentences and not just produce code. They are being made to reflect on the process, and are also given room at the end of each report for summarizing what they, personally, have learned as a result of this exercise. However, the author ends up spending an enormous amount of time, often 15 minutes per student, giving feedback[1].

When teachers request that students write papers on a specific topic, it can be instructional to first visit the site of a term paper mill to get an idea of the papers that are available there. Using the search term "write my term paper" with a search engine will also show the companies that are willing to custom-write a paper for around $10 to $15 per page. However, not all papers sold as original are indeed unique, especially not when they are cheap. For a well-written and well-researched paper, one can expect to pay $35 to $50 a page. Educators can at least drive the price of purchasing a term paper up by specifying in the assignment to the student that a reference to a particular, newly published, source must be incorporated into the paper. Or a comparison of a current event to similar events in the past can be part of the assignment, making it harder to find a paper mill with such a term paper on file.

Educators should be skeptical when students suggest a particularly narrow topic – they may have a friend who has just submitted such a paper to another university in town, or they may have found a copy on file at the student union or student club,

---

[1] With a class of 40 students, which is a typical class size at a German polytechnic, this means about 10 hours of grading per week, while only two class hours and two hours of preparation time are counted for an exercise session in the teaching load per week. Professors at German universities of applied sciences have 18 class hours per week to teach every semester. For universities with larger class sizes and very few teaching assistants or mid-level personnel, this is very costly, no matter how useful regular feedback is from a didactical perspective.

or even in the library. Discuss the topic with the student – they may have a genuine interest in this particular aspect – and see if a twist can be added to make the topic unique.

The students should not just hand in a term paper at the end of the term. Have them submit an outline, a first draft, and a commented bibliography. The latter should include naming the library where the material was found and giving the call numbers, or requiring them to record the search path that they took in order to find the materials online. This does, however, increase the correcting load tremendously for teachers, so it is only feasible for small classes. Certainly, where there is a demand, there is a company willing to fill that order. It is now possible, for an additional fee, to purchase a package with multiple drafts and even with an annotated bibliography, so even this method is not foolproof for keeping students from cheating.

5.1.5 Writing Courses and Clinics

There is a myth, in Germany at least, that students come to university well-prepared for writing and calculating. Having found that German students generally lack competence in mathematics, most universities offer bridging courses to help beginning students get up to speed. But writing skills are often sorely lacking as well. Especially in engineering disciplines or in medicine, students tend not to see the necessity for learning how to express oneself in writing. They do not understand the reasons behind academic writing and often equate the academic level of a paper with the number of footnotes or references it displays. Strangely, many are unaware that many books have been written in German about how to write at university level, and they have been available for many years. Manuel René Theisen's book on academic writing is currently in its 16th printing, for example, and includes an entire chapter on avoiding plagiarism (Theisen 2013).

Some students purchase term papers or hire ghostwriters or plagiarize material from the web because they just do not know how to write. They are afraid of the process, do not know how to read actively, do not know how to take notes, do not know how to structure what they are planning on writing, and do not know how to go about writing and revising. They often have the notion that one writes in press-ready prose, not understanding that texts may need to be rewritten many times to make them more understandable.

Many universities in the USA offer elective or even mandatory writing classes for their students. In Germany, this is seldom the case, although some universities are indeed starting to offer elective courses in writing. This seems to be particularly necessary for a generation of students who only write on a screen and often only regularly write 140 (tweet) resp. 160 (SMS) character messages, if at all.

It is important for educators to realize that this is also a generation that grew up with television, movies, Internet, and videos, and as such will tend to be more oriented toward finding information online. They are not accustomed to reading or finding information in non-searchable books. They may also have difficulty in critically examining online texts, as the sameness of the presentation will often mask the purposes that are behind the texts. Critical evaluation of Internet sources is, however, something that is usually not taught in schools at present, often because the teachers themselves are unsure about their own abilities to determine the quality of online sources.

Offering introductory courses in writing and setting up a writing clinic can be especially useful for students who are acutely struggling with a paper or thesis, as well as for foreign students who are still unsure when writing in the language of their host country. Individualized help can be given, and students from cultures that do not consider plagiarism to be a problem can be made aware of what is expected from them. It may be hard for German students to ask for help, so if such a clinic is set up, there will have to be some effort invested in making it easy for people to drop in anonymously and without an appointment and get help. Getting the word out that such a service is available will also take some additional work. Research institutions that employ researchers who struggle with writing publications in a language that is not their native tongue could also profit from a local writing clinic. The editors who work in such a clinic could contribute to more understandable publications, while understanding the fine line between making helpful suggestions, doing light editing, and ghostwriting.

5.1.6 Understanding Quotation

Many students appear not to understand the point of quoting the works of others in academic writing. Footnotes are seen as a quantitative property of a thesis and demonstrate just that the writer purports to have read widely on the topic. While this is certainly the case that it is vital to show what others in the field have done previously in the area of inquiry, it is also important that students and researchers understand the necessity for good quotations and footnotes.

There is also confusion over the difference between a citation and a reference, between indirect and direct quotations, and the need for footnoting in general. Many students hold the general impression that citations are extremely complicated to deal with, and if you are missing a period after the journal article name or did not put the year in bracket then you are also guilty of plagiarism and will fail.

There is a very simple rule about quoting: If the words you use are not from you, they need to be quoted so that the reader knows what is from the author and what is from someone else. And it is quoted by clearly marking the beginning and the end of the portion used, and giving a citation referring to an entry in the reference section where one can go to verify the veracity of that which is quoted.

Proper quoting is not just a special terror that has been developed in order to torture students. It is also a service that is offered to the reader, who might be looking for additional writing on the topic; they might be looking from the vantage point of a different question asked to the same material; or the reader might even violently disagree with what is being referenced and need to find the source in order to set up a proper refutation.

This is a different matter from sourcing statements of fact. There are many facts in each area if investigation that are considered to be "known" and thus not in need of referencing. The hard part is knowing what can be considered known and what is specialized knowledge. However, any kind of statement of statistics will need a reference to the source, as will little-known facts. The citation itself must be as close as possible to what is being referenced. It is never sufficient to only list a source in a bibliography, as it is not feasible for a reader to be forced to check every reference in the hopes of somewhere finding the source for a particular statement.

When exactly does something have to be referenced? Kate Williams and Jude Carroll (2009, pp. 26–27) give a good overview in their little handbook for students:

You need to reference when you:

- use facts, figures or specific details you pick from somewhere to support a point you're making – you **report**
- use a framework or model another author has devised. Let's say you '**acknowledge**'
- use the exact words of your source – you **quote**
- restate in your own words a specific point, finding or argument an author has made – you **paraphrase**
- sum up in a phrase or a few sentences a whole article or chapter, a key finding/conclusion, or a section – you **summarise**.

You don't need to reference if you:

- believe that what you are writing is widely known and accepted by all as 'fact'. This is usually called 'common knowledge'
- can honestly say, 'I didn't have to research anything to know that!'.

But
If finding it out did take effort, show the reader the research you did by referencing it.

In any case, it is necessary to precisely delineate what is taken from other sources and what is from the author, as discussed in Sect. 5.1.7, and to give exact references to make it easier for others to find and verify the sources. Phillip Theisohn (2012, p. 102) points out the difference between earnest and "camouflage" citation:

Das, was [der Plagiator] nicht weiß, vergessen, verlernt oder nie begriffen hat, sind keineswegs die Regeln wissenschaftlicher Arbeit, sondern der *Zweck* dieser Regeln, das ethische Fundament, auf dem sie errichtet wurden. Und so ist es ihm durchaus möglich, ein paar Hundert Seiten Text zusammenzustellen, die unter formalen Gesichtspunkten der akademischen Eigentumsordnung Genüge leisten, obgleich sie in Ansehung der Aufgabe, die Wissenschaft eigentlich zu bewältigen hat, völlig versagen. Man orientiert sich am Wortlaut, der nicht übereinstimmen darf, und setzt Fußnoten nur, um sich im Zweifelsfall gegenüber Plagiatsvorwürfen abzusichern. Um es offen zu sagen: Wir haben es nicht mit Wissenschaft, sondern mit wissenschaftlicher Camouflage zu tun. Und wenn unser einziges Kriterium in Fragen der geistigen Ökonomie die Einhaltung der Formalia ist, dann urteilen wir im Zweifel nur darüber, wie gut das Wissenschaftskostüm sitzt, das sich eine unwissenschaftliche Arbeit übergestülpt hat.

What the plagiarist does not know or has forgotten or unlearned or just does not understand is not the rules of academic writing, but the *reasons* for these rules, their ethical foundation. It is, of course, possible to put together hundreds of pages that follow the letter of the law of academic civil life, although they completely miss the point of what academic endeavor is trying to accomplish. They are only concerned with avoiding a word-for-word correspondence, and only use footnotes in order to avoid accusations of plagiarism. To be perfectly frank: we are not dealing with academics here, but with academic camouflage. And when our sole criterion for questions of intellectual business deals with the conformance to such formal aspects, then we are only judging about how well the academic costume fits that such an unscientific work has wiggled into.
(translation by the author)

A disregard for proper referencing is being passed on from one generation of academics to the next since there are plagiarists at all levels of the university who are replicating in their students what they have forgotten or unlearned or never understood. It is vital for academic inquiry that sources are able to be questioned and checked in order to determine that they do, indeed, say what they are quoted as stating. The next section will give more detail on this topic.

5.1.7 Proper Quoting

In order to make the difference between plagiarism and transparent quotation clear, it is necessary to have a coherent view of how the words or ideas of others can best be incorporated into one's own work. This includes questions of demarcation of the material and proper citation and referencing of the source, which will be discussed below.

5.1.7.1 Demarcation

In order to make evident exactly what part of a text is from the author and what part is from someone else, it is important to precisely mark the beginning and the end of the material that is taken from others. This includes both the textual replication of words or the paraphrase of the ideas of others. These are often referred to as direct or indirect quotations.

When replicating the words from someone else, the demarcation is easy: Quotation marks are used to begin and end the text so used, or the text is set off by an indentation. But inherent in the word "replication" is the notion of a copy, and for a direct quote to be true it must indeed be an exact copy. If parts of the passage must be left out for some reason, an ellipsis ("[...]") is used to denote that something is missing. If spelling or grammar must be adjusted so that it fits the flow of the sentence, this, too, is noted in square brackets ("[s]he" or "[the author]"). And if there in an incongruence or a misspelling in the original source, inserting "[sic]" (a short form for *sic erat scriptum*, "thus was it written") into the text at that position makes it clear that this is not a transcription error.

Demarcating indirect quotation is a bit more complicated. The introduction of the passage is easy, as the author is named: "As Smith states ...". What follows must *not* be an exact replication of the text (or else a direct quotation using quotation marks would need to be used). Instead, the ideas of the author quoted are paraphrased, that is, summarized in the words of the author using the quotation. It is not sufficient to just swap a word or two with a synonym, the grammar and the structure of the statement must be different from the source. Of course, it goes without saying that the paraphrase must render the ideas true to what the author intended.

The problem of indicating the end of the passage is much more difficult. Setting a citation mark, either in a footnote, an endnote, or an inline reference, is one of the easiest ways to note the transition back to one's own words. There are schemes that state that if the reference is given before the period at the end of a sentence that it only refers to the sentence itself, and if it comes after a period, it refers to the entire text from the beginning of the paragraph up until this point, but this is not universally agreed on. An often-seen method involves putting a citation only at the end of each paragraph, denoting where the ideas are taken from. However, if not only the ideas but also the words were taken, this method does not clearly delineate the transition.

The demarcation needs to be as close-fitting as possible, not somewhere in the middle of the passage and not including own material that now somehow seems to be in the words of someone else. Reusing the name of the author as a closing marker could also be useful for making the boundaries clear, for example as "[...] was also noted by Smith."

5.1.7.2 Citations and Referencing

It is important not to confuse the concepts of citations and references. *Citations* are given in footnotes, endnotes, or inline and give the exact location of the material used. A citation is the in-text marker giving the source of the material used and should be given as close to the demarcation as possible. The point of a citation is to allow the reader to easily find the source used.

References are listed in the bibliography or works cited at the end of a chapter or at the end of the thesis and give detailed information that is necessary in order to obtain the source. It is important for this information to be as exact as possible in order to permit the reader or their university librarian to quickly find the work in question. If a source needs to be obtained by interlibrary loan, sometimes only the pages in question from a journal will be photocopied, so giving a wrong page number can be costly for the reader both in time and page charges.

Often a short form for referring to this material in the citation is defined, such as a number or the author's name(s) and the year, or an abbreviation of the names and year. These schemes can differ widely from field to field, and some publications insist on using their own particular method. The short form used in the citation should, if being used specifically to source a quotation or fact, include details such as the relevant page number in order to make it easier for the reader to find the exact position of the fact or quotation cited. Citations can either be given in footnotes at the end of a page or in endnotes at the end of a chapter, or as a so-called "American" citation style. This method puts a reference right in the text, enclosed in brackets or parentheses, so that it does not interrupt the reading.

It is never sufficient to simply list the materials used in the bibliography. When the source for a particular statement needs to be found, it is quite an imposition to make a reader have to guess which of the references is a likely source, and then dig through that material looking for the correct page. This is not acceptable as it costs the readers much time. Writers need to strive for transparency as to exactly which statement comes from which particular source. This also makes further research into this area by the writer easier, as the steps needed to find the information do not have to be retraced.

Often citations will use a reference multiple times, referring either to the same page or a different page. There are Latin indicators that can be used as abbreviated citations, but only for footnotes or endnotes. Encountering one of these in an inline citation would cause much puzzlement for the reader as it would be necessary to leaf back through a number of pages, scanning for the matching citation.

Ibid.   (*ibidem*, means "the same place") for denoting something taken from the same reference; if it is from a different page, that page is given

Loc. cit.   (*loco citato*, means "in the place cited") for referring to something that is in the same work and on the same page as the previous reference, so it is never followed by a page number

Op. cit.   (*opere citato*, means "in the work cited") for referring to a previously cited work or one that can be found in the bibliography

Here is an example the author put together in order to illustrate the use of all three styles:

[1] Booker, A. (2001). *Life and the Universe*. New York: Fulmacron, p. 12.
[2] *Ibid.*
[3] *Ibid.*, p. 14.
[4] *Loc. cit.*
[5] Decker, C. (2006). *The Universe and Life*. London: Harbinger.
[6] *Ibid.*, recital 4.
[7] Booker, *op. cit.*, p. 42.

Citations 1 and 2 both refer to Booker, p. 12.
Citations 3 and 4 refer to Booker, p. 14.
Citation 7 refers to Booker p. 42.
Citation 5 is a general reference to the entire book by Decker.
Citation 6 refers to a specific marginal recital in Decker.
Citation 2 could also have been *Loc. cit.*

This manner of referencing is sometimes considered to be old-fashioned, as more and more people are adopting styles such as the *APA* (American Psychological Association) or *MLA* (Modern Language Association) style. This book, for example, uses a style that was adapted from the APA style. These suggest using an *author/date* citation system, which has the advantage of not forcing the reader to refer back to the references every time they want to know if a particular person or research group is being cited.

Internet references can also be made by including the URL and the date on which the URL was cited. The reason for giving the date is so that even if the page has been removed or the site reorganized since the reference was recorded, the *Internet Archive* may have a copy of the page stored. The goal, as with all references, is to make it easy for the reader to find the work in question. There are some services such as *WebCite* that will archive a copy of a web site; however, this particular service does have a financing problem and may not be available in the future.

Interestingly, some authors apparently copy the citations or references from a source, as well as the text, and use them as if they had read and analyzed the material themselves. This copying can be quite hard to determine, as it is entirely possible that they did read the exact same literature. But in the case in which there are errors in the bibliographic entries and these same errors occur in the later thesis, it is clear that they were just copied, not examined independently. Sometimes papers that never actually were printed or existed in the referenced form take on a life of their own, serving as a reference in paper after paper. See Dubin (2004) for an extensive example.

Secondary citation is used when the author wishes to refer to something, but only has a secondary source available and cannot access the original. While one should always strive to work from the original sources, sometimes this is just not possible. The transparent way to deal with this problem is to write "So White, 2005, in (Brown 2007)" or as "(Brown 2007) quotes (White 2005) as stating something."

Some writers expand their reference section with works that they may have consulted during the work on the thesis, but which were then not actually used in citations. These *garnish references* are sometimes included to make the writer appear to have used much more material than they did. While it is possible for a few of these references to slip through the cracks during the process of editing a text, having 10% or more of the entries in the reference section only being there to pad the list is not acceptable.

Unfortunately, there is a similar problem that arises when a journal editor explicitly requests that the author include a certain number of references to a particular person's work or to other work that has previously appeared in a particular journal. This is done to increase the impact factor of the journal, as has been demonstrated by (Wilhite & Fong 2012), among others. Although such so-called pleasing or *coercive citation* does happen, the reference section should only contain those materials that are actually referenced. Other material could be placed in a separate bibliography that gives a wider overview of the field of study and is perhaps also commented.

5.1.7.3 Reasons for referencing

Students are sometimes puzzled as to why references are even needed. They just "know" that something is a fact and are sure that this is some sort of universal truth. Some even see references as a quantitative indicator of quality. A "good" research paper needs to have a certain number of references, not too many, not too few. Because they have not formulated questions for themselves and tried to track down details that are not to be quickly found with a Google search, they tend to not understand what the reasons for referencing are.

Jude Carroll, an educational development consultant based in Oxford, England, has put together an overview of eight reasons for referencing material that she developed together with Kate Williams and uses in instructional materials in her workshops. They are listed in Table 5.1.

5.1.8 Reading and Checking

One of the more shocking realizations that has come about as the result of the discussions of plagiarism is that some German professors do not actually read very thoroughly what they are grading. Some professors apparently do not even read the theses at all, as cases that excessively use Wikipedia texts or even recycle material written by the graders have shown. Grades are then based either on the introduction and summary of the work, or on the oral presentation of the thesis, or the second examiner just agrees with the grade given by the first examiner in order to save time, faced with an ever mounting pile of written work needing attention.

It is possible that this is a reaction by university teachers to the phenomenon in Germany that the universities are being asked to educate more and more students, while the number of professors stagnates and the mid-level teaching personnel is being drastically cut (Michler 2011). While there used to be just the diploma thesis to grade, there is now a bachelor's thesis and a master's thesis to read and grade for each student, and university professors also have doctoral theses to read, in addition to all of the term papers. And since "good science" is often measured in the number of doctoral students one professor has, it is no longer seldom that one professor has 10 or 25 or even 100 doctoral students, especially in medical or natural sciences or in business studies, where external funding must be obtained.

Educators complain, of course, that they would be glad to read the proper term papers and theses if only a software could weed out the plagiarisms. But since the software does not work as desired, as explained in Sect. 4.1, and because it costs so much time to detect, document, and discipline plagiarists, some teachers just give a slightly worse grade if they have the feeling that the paper might be a plagiarism and continue with the next paper from the pile.

Universities must make it possible for their teaching staff to have time to read the papers, and it must become part of the school culture that the professors do read papers themselves and not just delegate this job to a graduate student. They also need to be trained in the art of spotting plagiarists and data manipulators and ghostwriting, and they need to have tools and procedures available in order to deal with suspicions and confirmed cases of plagiarism. These procedures should not be just codified, as is often the case with university rules for good academic practice. The procedures must be taught to new faculty, lived by all members of the university,

Table 5.1 Eight reasons for referencing material, adapted from (J. Carroll, private communication, 13 March 2013)

What the reference shows	Why we do it
Authority	We use other people's research and work to support our own assertions and give credibility to what we are saying. Referencing is very closely related to the need in academic work to provide evidence for any assertions.
Collegiality and politeness	Academic knowledge is a collaborative project of individuals. It is important to acknowledge the individual efforts and contribution to the research of all participants.
Traceability	When academics develop a theory or idea, they build on other people's work and they acknowledge this by referencing. A reference means it is possible to check if an author has used other researcher's work correctly.
Validity	A researcher's contribution can be evaluated by looking at the sources they have used – have they chosen key sources? Have they used up-to-date sources?
Accuracy	Referencing is essential when defining key terms such as market segmentation, schizophrenia or post-modern, etc. One needs to demonstrate that a recognized definition is being used. This is generally not the Wikipedia definition, but there may be a good reference to a source to be found there.
Breadth of research	The references in a paper or thesis show the reader what has been read and will influence the reader's impression of the author.
Scholarship	Through referencing, an author shows what has been read and demonstrates that the issue under discussion has been thoroughly researched.
Protection against accusations of plagiarism	With proper citation, others cannot say that the author took others' work without proper acknowledgment.

and enforced across the board. This may make it necessary for universities to require faculty to report suspected cases of plagiarism.

5.1.9 Talking about Plagiarism

A discussion about good academic practice and how to avoid plagiarism needs to be initiated at every institution of higher learning. It cannot be assumed that the students magically know what this is, even though it would certainly be helpful if this were taught in the secondary schools. Most particularly, it is necessary for educators to discuss the use of the Internet and Internet-based resources such as the Wikipedia with their students.

It is also necessary for universities to be determining how to assess work that is done collaboratively. Graduates are expected to be good team members, and to be able to work together to produce results. This often lies in direct opposition to questions of sole authorship, as would be expected for a thesis. Effort must be expended in order to develop and communicate ways of marking the contributions of the individuals in collective works. There are not always clear ways of doing so, thus making it imperative for the active participants in academic endeavor to discuss with each other how best to do this.

Teachers at secondary schools and professors and instructors at university need to be trained in how to find plagiarism and how to avoid it, much in the way that ethics education is required in the USA for researchers. It does not need to be an intensive training, but a day's course at the beginning of a teaching career and a few refresher sessions, mostly to discuss new ideas and methods with others, should be offered by the schools and universities.

Universities in Berlin have been offering new professors a teaching load reduction in their first two semesters if they participate in didactical training. This includes discussions about plagiarism and how to deal with the problem. New professors are said to find the didactical training quite useful, particularly if they have not previously had formal training in education.

5.1.10 Swearing an Oath?

A common solution against plagiarism in Germany is having students swear an oath, on their honor, that they have properly marked all material used and not falsified any data in their research. The reason for this is a purely legal one: It is easier to take the student to court because they will have committed perjury if it can be shown that they used unmarked material or falsified their data anyway. Using unmarked material that is no longer under copyright does not have legal consequences, but it is not acceptable in research.

In addition, if a university rescinds a degree or refuses to award one because of academic misconduct and the student takes the university to court, being able to demonstrate that the student swore a false oath will strengthen the case of the university. For some reason, there is a widespread feeling that swearing an oath in some manner deters bad academic practice, as it is sometimes offered as a solution to the problem of plagiarism and is included in many university by-laws, even though it is often not legal for the universities to insist on the students swearing such an oath.

But the question of good academic practice should rather be one about ethics and morals and not just one of legalities. It is hard to precisely determine where to draw the line when it comes to bad academic practice, thus the necessity for discussing the topic and imparting a moral sense of what is the right thing to do to the students. If swearing the oath comes at the end of some formal ethical training on the topic, it might be useful. On the other hand, if it is just another signature to be done before handing in a thesis, it will probably not have much effect.

5.1.11 A Sense of Honor

Instilling a sense of "we don't plagiarize here" is, according to Donald L. McCabe, one of the best ways to keep students from plagiarizing. His work at the *International Center for Academic Integrity*, located at Clemson University, has found that universities that have honor codes and insist that their first-year students formally pledge to follow the honor code have significantly less problems with cheating than universities without one (McCabe & Treviño 1993; McCabe, Treviño, & Butterfield 1999, 2001).

Callahan (2010) reports on a number of studies, and specifically cites a 1964 study, as well as the McCabe and Treviño study (1993) which confirmed the older paper. McCabe, Treviño, & Butterfield looked at the honor codes again in (1999). Interestingly enough, although the honor code schools had less problems with cheaters, one of the lowest levels of cheating was found in a school without an honor code, and one of the highest at a school with an honor code, they report. So an honor code is not an automatic measure to reduce cheating.

Looking carefully into their data, McCabe, Treviño, and Butterfield (2001, p. 224) found that

> [...] although this noncode school did not have a formal honor code, it had developed a culture that emphasized many of the elements found at code schools and encouraged academic integrity without instituting a formal code. At this school, administrators and faculty clearly conveyed their beliefs about the seriousness of cheating, communicated expectations regarding high standards of integrity, and encouraged students to know and abide by rules of proper conduct. In contrast, the honor code school, although it had a 100-year-old honor code tradition, failed to adequately communicate the essence of its code to students and to indoctrinate them into the campus culture.

It is not enough to just have an honor code and go through the motions. A pledging ritual can be quite elaborately celebrated at some American universities and involve speeches, singing, and the act of signing (Vanderbilt University n.d.) or even a church service that includes the ritual of touching an original cornerstone of the first university building (Seewanee University of the South 2012). Such a ceremony can surely contribute to internalization of a culture of quoting, but the faculty also needs to live the honor code, thereby setting a good example for the students. Universities must communicate what the honor code is about, and see to it that people who break the rules are adequately and transparently dealt with.

McCabe also finds it vital to have student involvement in developing the honor code. This gives them ownership of this aspect of ethical community building, and keeps it real, alive, and with personal meaning to the students. He notes that failure on the part of the university and the faculty to act on the question of ethics and honor sends a powerful message to the students: that this is not a priority question.

The ICAI has put together a list of some universities that have honor codes and link to the texts and to the ceremonies that are used (International Center for Academic Integrity n.d.). They also have an article by Pavela (1997) available on their site with an example honor code, as well as much other material dealing with fostering academic integrity at university. Appendix F gives some more details about the ICAI.

Since Germany understandably has quite a problem with formal rituals and public and collective oath swearing stemming from their massive misuse during the 1930s and 1940s, it might make sense to adapt the notion of honor code to entail just signing a pledge during orientation and not overdoing the ceremonious rituals.

5.2 Transparency

One of the most problematic aspects of dealing with plagiarism is the secrecy that shrouds many of the proceedings. The tenets of good academic practice should be easily discoverable and spoken about often. If accusations of plagiarism are made, the processes themselves need to be documented and publicized, and the results – especially those that involve a revocation of a degree – need to be appropriately communicated.

Mandatory publication of dissertations, as has been the norm in Germany since the late 19th century, is an excellent tool for helping identify persons who, indeed, hold doctorates and for permitting others to examine the material in order to determine if plagiarism is involved. This subchapter will address these aspects in more detail.

5.2.1 Policy on Academic Practice

It is important that each university has a policy on good scientific and academic practice that does not just define the crimes and list the punishments – the goal of the policy should be to delineate between good and bad academic practice so that students may learn and understand how to do the one by avoiding the other. The procedures to be followed in cases for which an allegation of bad academic practice is made should be clearly defined in such a document.

It is important that the contents of this document be communicated throughout the university. It is not enough to just post a legalese document somewhere deep on the university web site and link to it from some obscure page. Students and faculty alike need to be made aware that the policy exists and popular student myths about what is permitted and what is not permitted in good academic practice need to be dispelled.

It is not necessary to start such a document from scratch. Many good examples can be found, especially on university web sites in the USA. These can be used as inspirations for setting up an own document. If a university decides to use a portion of a policy verbatim, it would be important to ask for permission to copy and to make clear which part has been copied. It would be quite embarrassing to have a document on good academic practice itself be copied without attribution.

5.2.2 Honor Code and Board

Having some sort of body that is given the task of evaluating charges of bad academic practice and suggesting sanctions to be imposed takes the pressure off of individual faculty, who otherwise end up having to evaluate the situation as well as decide on and mete out sanctions.

There are variations on the constitution of such a board. The members can be appointed or elected. There can be a majority of faculty or students or a parity situation. They can be charged with investigating cases or they can just decide on cases that others have prepared and submitted to them. The university can make it mandatory for faculty to refer all cases of academic misconduct to the board, or it can reserve judgment only for more serious breaches. If there are many cases that come up, a large university might decide to have departmental honor boards, especially as in Germany doctorates are granted on that level[2]. There is no ideal structure; much will depend on the traditions at a particular university and how well the board is made known to the local academic community.

The honor board should, however, be a body that meets regularly and not just when a case is submitted for consideration. The university needs to publicize the

---

[2] The *Fakultät* is the level of a collection of departments somewhat akin to a "School" at an American university, which decides whether to confer a doctorate or not. It is the university, however, that actually awards the degree.

fact that such a board exists, and should set up procedures for whistleblowers to report on misconduct without having to reveal their identities. For example, there can be an anonymous dropbox set up. It is important to encourage people to inform the honor board non-anonymously, but care should be taken to protect the identities of the persons stepping forward.

In any case, the honor board will need to have resources available for investigating, reporting, and documenting their work. If the university policy includes a provision for harder sanctions for repeat offenders, then it will be necessary for there to be an effective system of keeping track of violations. Neufeld & Dianda (2007) note that for a second offense, some universities suspend a student for at least a semester as a minimum penalty. Others are willing to expel a student for getting caught a second time. However, it will be necessary to make sure that the legal basis for such a suspension is given, especially for publicly funded institutions.

McCabe & Pavela (2000) summarize their experiences from years of investigating cheating and honor codes at universities in the USA. They have empirically demonstrated that introducing an honor code led to reduced rates of self-reported cheating. They make it clear, however, that while honor codes can be useful in fostering a culture of good academic practice, they do not guarantee honesty just because they are in place. The university administration must be committed to honesty and communicate this goal while supporting the board. Appendix E gives more information on setting up an honor code and board.

5.2.3 Permanent Record

Neufeld & Dianda (2007) noted that some method is needed for recording academic misconduct on transcripts. The diploma supplement, which accompanies a transcript in the EU nowadays, would be the ideal place to record this. They recommend a two-step process: internal records keeping for a first offense, but that those found to have been cheating a second time need to have some notice of this on their permanent record and transcript. There might be a defined process, for example, successfully completing a course on academic integrity, that could lead to removing this mark from the permanent record, or information on cheating behavior at university could be erased upon successful completion of the degree.

But doctoral degrees that have been granted and then rescinded pose special problems. Even if universities decide that a case of academic misconduct is serious enough to warrant taking back the degree, the universities are not always able to contact the person, especially if they are now in a foreign country. The only time that it is usually necessary to produce the doctoral certificate will be when having the title put on an identity card in Germany, or when joining a new company. Perhaps a central, searchable, online database of doctorates needs to be organized that keeps track of persons having doctorates granted from foreign countries and doctorates known to be rescinded. It would make sense to keep such a register in the national library.

5.2.4 Open Science

Good academic practice is a question that needs to be discussed within the academic community, and it needs to be continually and openly discussed. Transparency is the key to keeping people honest. If the materials that one writes are open to objective criticism from the outside, and if the discussion of academic misconduct is also done in public, this social pressure will greatly reduce the temptation to lift other's words. If the data collected from a scientific endeavor is recorded in public and then the conclusions are drawn from the data, it is not possible to exclude data points that are uncomfortable.

Of course, publishing collected data will mean that others can also draw conclusions from the data. This means that the system by which we award recognition to scientists – the publication of papers – will need to be adjusted accordingly.

5.2.5 Amending the Public Record

After an academic work such as a dissertation has been determined to be a plagiarism and a title revoked, there are some interesting questions that arise. It is clear that the publisher will not be selling more copies of the book, but do they have to try and call back all of the copies already sold? Is the book to be removed from the bookshelves of the libraries because it is a plagiarism? Should it only be stamped "plagiarism" with an appropriate notice in the card catalog and the OPAC? How is it to be noted in the OPAC, should the notice that it was a dissertation be removed, or should an additional notice record that the doctorate was rescinded? What needs to be done if the book has already been quoted by others? Eric Steinhauer (2011) blogged about some of these questions in connection with the zu Guttenberg case.

There are more questions: Who makes sure that a person removes the "Dr." from the official state identity card when the doctorate has been rescinded? Who makes him or her take it off their office door and their stationary and their web pages? This is actually a similar question to knowing if someone using a doctorate actually has one. There is no official instance charged with making sure that a plagiarist stops using a rescinded doctorate, although Germany does have laws such as § 132a Strafgesetzbuch, *Mißbrauch von Titeln, Berufsbezeichnungen und Abzeichen*, that forbid misrepresenting oneself as to having a doctorate if this is not the case. In other countries, for example, Switzerland, it is not forbidden to use a false title, but universities do not like being misled and will see to it that the person in question is no longer associated with the university, as happened in a recent case reported in (Stäuble 2013).

But in order for legal action to be taken, a complaint must be lodged with the authorities. And since foreign doctorates are usable in Germany without the thesis being registered in the national library, there is no way for someone to independently determine if someone using a doctorate actually has one, other than requesting to see the original of the degree certificate, as would be necessary when commencing employment as a university lecturer or researcher. Since people are understandably hesitant to accuse anyone of misrepresenting him- or herself without evidence, it is easy to continue using a doctorate even if it was revoked. It is unknown if this is at all widespread.

For plagiarisms that are published in academic journals, there is the notion of a published *retraction*. After the plagiarism has been substantiated, a printed journal will publish a short notice that this or that article has been retracted, often giving a brief reason and stating whether the author or authors agree. The blog *Retraction Watch* often comments on the retractions that it finds, usually in the medical literature.

Retractions of papers published online are usually handled by replacing the paper with a notice of retraction. During one of the VroniPlag Wiki investigations it was found that an author, *Nm*, had not only plagiarized widely in his dissertation, but many of the papers and conference proceedings in his publication list were also plagiarized. The publishers were informed and a list was started online at VroniPlag Wiki (2012). To date only eight of the papers have been formally found in violation of publication principles and withdrawn. The rest are still pending as the publisher in question had no formal procedure for withdrawing conference proceeding contributions.

One of the major obstacles has turned out to be that there is no easy way to publish a retraction for conference proceedings as they are not always published as a series. Each volume stands alone even if the conference takes place regularly. Journals generally have a next issue; conferences do not. And with the multitude of mock conferences springing up with self-published conference proceedings, this question of how to publish a retraction for a conference paper remains unsolved. Perhaps they should not be included in CVs for exactly this reason.

5.2.6 Public Reporting of Infractions

Neufeld & Dianda (2007, Public Reporting Requirements) also suggest that the public reporting of infractions could serve as a performance indication – not just in counting infractions, but in "demonstrating the seriousness with which it treats instances of academic dishonesty among its students." The Office of Research Integrity in the USA publishes the names of researchers found to have demonstrated bad scientific practice for a set number of years on their homepage.

By 1865 in Germany, as shown in Sect. 3.5.1, the names of plagiarists were already being published in academic journals. In the present day, the names of some authors of dissertations and habilitations with extensive, documented plagiarism have been published on the VroniPlag Wiki home page. There has been extensive debate on the topic of whether putting people's names on the Internet is akin to the medieval practice of placing delinquents in the stocks in public and whether such public shaming is ethical practice.

A monograph about medieval law by Wolfgang Schild notes that, in order to understand what the punishments of the time meant, one must understand the concept of honor (Schild 2010, p. 182). From around the 12th century, Schild writes, a person was identified by his place in society, there was no concept of the individual. Any change of status changed the person.

Schild describes the punishment of putting someone in the stocks as not being meted out by officials, but by society as a whole, by means of the ridicule to which the delinquent was subjected. This also had the additional effect of giving the general public a good feeling of belonging – we are the good guys, ganging up on this miscreant. And they quite enjoyed the spectacle, Schild states.

Is publishing the name of someone on the Internet together with some documentation of a misdeed the same as putting them on the stocks? It is now possible for them to be easily found by others using a search engine and quickly connected to their wrongdoing. But since information can remain on the Internet for a very long time, it is important that one is completely sure that the person named is responsible for his or her actions before naming them. Anyone named in error will have a hard time clearing their name.

Should the shaming be forever? Criminals who have served their sentences are given a chance for a fresh start in life. Or is bad academic practice so serious that a wrongdoer must be marked for life? VroniPlag Wiki has even had people request that their names be removed from the site, although they are only mentioned because they wrote a review of the work in question or their works were plagiarized by others. If the plagiarized works are not permanently marked, however, there is a danger of a future academic quoting from them. A publication is something that is permanent, so an author needs to be very clear about the fact that publishing a plagiarism will incur a permanent shaming, should the plagiarism be found out.

It is not possible at this point to fundamentally change the Internet, but it is possible to change how we use the Internet and how we react to finding references on the Internet. It is important, in the opinion of this author, that published books that have been determined to be plagiarisms not be secretly removed from the stacks of the libraries. They need to be clearly marked in the catalogs and in the books themselves that they contain plagiarism. This will serve as a warning to those who would want to use the material that they will have to dig deeper in order to use the original sources. This puts the major focus on the material itself and not the authors. Bad science cannot be permitted to remain in the public record.

Is it important to publish the names of people found to have committed academic misconduct on the Internet? If everything is kept secret, they may carry on plagiarizing, or moving from job to job without the new employers being aware of these serious issues from the past. It is important for hiring committees to consult a search engine in order to investigate an applicant, but it is also important for them to temper their search by actually reading what they find to see if it is relevant, determining the date it was published to see if it is current, and verifying the veracity of the accusation to see if it needs to be taken into consideration. The plagiarists are given a second chance, not by having their plagiarism removed from public sight, but by publishing again and again without plagiarizing.

An additional hope for publicly naming plagiarists is to prevent others from following suit. It does seem that the absence of any form of effective punishment for plagiarism or academic misconduct may have led to the situation we are seeing today. But in general, deterrents do not work all that well. So the major goal of public reporting of infractions must be to give future employers a chance to find out about the problems from the past. Perhaps a central database in which those who have been depromoted are listed could be useful here. This would, however, place quite a legal burden on the keeper of such a database, as named plagiarists can be litigious as they try everything possible to keep their names off of such lists.

Should plagiarizing students be publicized? Here the author suggests rather not – they are beginners, still learning how to work properly. Situations in which students are found to be plagiarizing are excellent teaching moments. An instructor has their undivided attention, so perhaps the situation can be used to teach a useful lesson. And it also shows us as teachers where we need to apply more instruction (Hunt 2002). However, if students are caught cheating multiple times in some form or another, there should be some sort of publicity involved, perhaps even noting on their diploma certificates that they were involved in plagiarism.

There are no simple and easy answers to this question – but it is one that needs urgent attention and discussion.

5.3 Testing

Term papers and theses need to be checked if there is a suspicion of plagiarism. Who should be responsible for this? And is it necessary or desirable to have a central testing agency?

5.3.1 Local Responsibility

As will be seen in Chap. 6, in many countries the university libraries are charged with educating faculty and students about plagiarism. They are usually also active in selecting any software systems that might be used at the university, and they are often instrumental in developing a policy for dealing with plagiarism and informing the students about the policy.

Librarians are used to dealing with and organizing large and complex informational structures, and so they are also in a good position to understand and interpret the at times cryptic results of plagiarism detection software and will have a good overview of the databases that are available to assist in finding potential sources.

Library personnel can be seen as consultants who can offer advice as to how to go about detecting plagiarism, if a member of the teaching staff has a suspicion. They can help guide an online search, use and interpret plagiarism detection software, and assist in obtaining obscure materials that may have been used as a basis for copying.

The final decision as to the severity of the plagiarism found should, however, reside with the faculty member who has requested that plagiarism detection software be used.

If a university should decide to invest in a software system that checks all student papers submitted, then it would be important for there to be an institution such as a plagiarism consultant that a faculty member can meet with to discuss borderline cases. It will be very important for the library to offer training to faculty so that they understand that the software is not infallible. False negatives are irritating, because someone is getting away with plagiarism, and it is problematic to be relying only on the numeric results of the systems. False positives, however, are much more problematic, as an innocent student will be accused of cheating. Since university librarians tend to be permanent employees, they are in a good position to understand these problems and are able to educate new faculty about the careful use of such systems.

Setting up a plagiarism resource center in a library leaves the responsibility for dealing with plagiarism with the individual educator, but does not leave them floundering on the wide-open seas of the Internet without guidance.

5.3.2 Centralized Investigation?

Since cheating in general and plagiarism in dissertations in particular have been demonstrated to be a widespread problem in Germany and elsewhere, perhaps it is time to set up an external quality assurance investigation agency. Such an agency would be charged with taking a scientifically valid random sample of theses accepted every year and investigate them for plagiarism or data manipulation. The results would then be given back to the universities to deal with, which might involve both sanctions for the individuals caught out and setting up a process for avoiding similar problems in the future. The agency would have no authority for deciding on sanctions, although they could advise the university on possible actions to be taken. The final decision in any case must rest with the university and the department.

Testing a random sampling of theses would perhaps not only serve to inhibit persons from submitting plagiarized theses or duplicate publication of research materials, but it would also provide a quality indicator as to how much progress is being made in this area. Additionally, this does not single out the individual student, but the student, the advisor, and the department are being examined together.

There is, understandably, fear of such a central agency in Germany because the responsibility for universities lies solely with the individual states. However, the point of a central agency would be that it can operate independent of any local processes and contexts and would not be under any conflicts of allegiance. Such an agency would need to be funded by the federal ministry in order for it to maintain its independence from the individual states and universities.

Setting up such a national investigation office would be relatively expensive, as it is a time-consuming process to check individual theses for plagiarism without knowing the literature or having a first suspicion for possible sources. And with over 100 universities in Germany that grant doctorates, even obtaining an overview of the material to be investigated would be a daunting task. But seeing the current state of widespread and varied misconduct throughout Germany, as documented in part by the VroniPlag Wiki, it would be one useful measure to help improve the quality of the research. Identifying problem zones permits the universities to focus their efforts at raising the quality of dissertations to those areas that are most in need of them.

Study programs are re-accredited every five years in Germany, but there is no external accreditation of research policies or the quality of research, publications, or dissertations being produced at German universities. Measuring, if it is done at all, tends to extend only to counting the number of publications produced, the number of students graduated, or the amount of external funding acquired. Perhaps it is time to include an examination of the quality of research into the accreditation procedure, or even to go to the extreme of only permitting university faculties that pass an external accreditation every five to ten years to continue to award doctorates.

References

Bassett, T. J. & Porter, P. W. (1991). 'From the Best Authorities': The Mountains of Kong in the Cartography of West Africa. In: *The Journal of African History*, Vol. 32, No. 3, pp. 367–413. Available at http://www.jstor.org/stable/182661 cited 5 July 2013.

Callahan, D. (2010). Why Honor Codes Reduce Student Cheating. In: *Huffington Post*, 14 December. Available at http://www.huffingtonpost.com/david-callahan/why-honor-codes-reduce-st_b_795898.html cited 15 March 2013.

Dubin, D. (2004). The Most Influential Paper Gerard Salton Never Wrote. In: *Library Trends*, Vol. 52, No. 4, pp. 748–764.

Franklyn-Stokes, A. & Newstead, S. E. (1995). Undergraduate Cheating: Who does what and why? In: *Studies in Higher Education*, Vol. 20, No. 2, pp. 159–172.

Hunt, R. (2002). *Four Reasons to be Happy about Internet Plagiarism*. [Web page]. http://www.stu.ca/~hunt/4reasons.htm cited 22 August 2013.

International Center for Academic Integrity. (n.d.). *Educational Resources*. [Web page]. http://www.academicintegrity.org/icai/resources-4.php cited 3 July 2013.

Library of Congress. (2010). *American Treasures of the Library of Congress: California as an Island*. http://www.loc.gov/exhibits/treasures/trm101.html cited 5 July 2013.

McCabe, D. L. & Pavela, G. R. (2000). Some Good News about Academic Integrity. In: *Change*. Vol. 32, No. 5, pp. 32–38.

McCabe, D. L. & Treviño, L. K. (1993). Academic Dishonesty: Honor Codes and Other Contextual Influences. In: *The Journal of Higher Education*, Vol. 64, No. 5, pp. 522–538.

McCabe, D. L., Treviño, L. K., & Butterfield, K. D. (1999). Academic Integrity in Honor Code and Non-Honor Code Environments: A Qualitative Investigation. In: *Journal of Higher Education*, Vol. 70, No. 2, pp. 211–234.

McCabe, D. L., Treviño, L. K., & Butterfield, K. D. (2001). Cheating in Academic Institutions: A Decade of Research. In: *Ethics & Behavior*, Vol. 11, No. 3, pp. 219–232.

Michler, I. (2011). Professoren leiden unter Ansturm der Studenten. In: *Die Welt Online*, 8 August. http://www.welt.de/dieweltbewegen/article13532016/ Professoren-leiden-unter-Ansturm-der-Studenten.html cited 5 July 2013.

Neufeld, J. & Dianda, J. (2007). *Academic Dishonesty: A Survey of Policies and Procedures at Ontario Universities*. [Working paper]. Available at http://www.cou.on.ca/publications/academic-colleague-papers/pdfs/academic-dishonesty-a-survey-of-policies-and-proce cited 10 May 2013.

Pavela, G. (1997). Applying the Power of Association on Campus: A Model Code of Academic Integrity. In: *Journal of College and University Law*, Vol. 24, No. 1, pp. 97–118. Available at http://www.academicintegrity.org/icai/assets/model_ code.pdf cited 22 August 2013.

Risko, E. F., Buchanan, D., Medimorec, S., & Kingstone, A. (2013). Everyday attention: Mind wandering and computer use during lectures. In: *Computers & Education*, Vol. 68, pp. 275–283. http://www.sciencedirect.com/science/article/pii/ S0360131513001218 cited 18 August 2013.

Schild, W. (2010). *Folter, Pranger, Scheiterhaufen: Rechtsprechung im Mittelalter*. München: Bassermann.

Seewanee University of the South. (2012). *Honor Code Signing*. [Video]. http:// www.youtube.com/watch?v=5S6u0Q_Sbwk cited 22 August 2013.

Stäuble, M. (2013). Der Scheindoktor. In: *Tages-Anzeiger*, 9 August. Available at http://www.tagesanzeiger.ch/zuerich/Der-Scheindoktor/story/26179683 cited 9 August 2013.

Steinhauer, E. (2011). Guttenberg aussondern? In: *Bibliotheksrecht*. [Blog], 2 March. http://www.bibliotheksrecht.de/2011/03/02/guttenberg-aussondern-10740355/ cited 1 January 2013.

Theisen, M. R. (2013). *Wissenschaftliches Arbeiten: Erfolgreich bei Bachelor- und Masterarbeit*. 16th ed. München: Vahlen.

Theisohn, P. (2012). *Literarisches Eigentum: Zur Ethik geistiger Arbeit im digitalen Zeitalter. Essay*. Stuttgart: Kröner.

Vanderbilt University. (n.d.). *Honor Code Signing Ceremony*. [Web video]. http: //blip.tv/studentvu/honor-code-signing-ceremony-68584 cited 3 July 2013.

VroniPlag Wiki. (2012). *Nm/Comparisons*. [Web page]. http://de.vroniplag.wikia. com/wiki/Nm/Comparisons cited 22 August 2013.

Wilhite, A. W. & Fong, E. A. (2012). Coercive Citation in Academic Publishing. In: *Science*, Vol. 335, No. 6068, 3 February, pp. 542–543.

Williams, K. & Carroll, J. (2009). *Referencing & Understanding Plagiarism*. Basingstoke: Palgrave Macmillian.

Wilson, F. & Ippolito, K. (2007). Working Together to Educate Students. In: T. S. Roberts (Ed.), *Student Plagiarism in an Online World: Problems and Solutions.* Hershey, PA: IGI Global.

Chapter 6
Plagiarism Policies and Procedures in Other Countries

An instructor has determined that a student paper or thesis is a plagiarism. What happens now? There is no simple answer to that question. Is this a first-year student? Then it might be possible to turn the situation into a teaching moment. The instructor has the undivided attention of the student, and can teach about good academic practice. Is this a final thesis, or even a dissertation? That is unacceptable. The exact procedures are different from institution to institution – and some universities in Germany have no defined procedures whatsoever.

Possible sanctions will depend on the importance of the assignment plagiarized and the status of the student. Examples are

- oral warnings,
- written warnings that are included in a student's permanent record,
- a grade reduction that may include reducing the grade for the assignment to zero,
- possible additional remedial work such as writing an additional paper about plagiarism or having to solve additional exercises,
- remedial work with a capped grade,
- a failing grade for the course,
- suspension for a limited period of time,
- expulsion from the program, or
- expulsion from the school.

Should the deceit only be discovered after a degree has been awarded, then the degree can even be rescinded.

A major question is one of policy – does the university have a published policy on plagiarism that outlines the penalties that academic misconduct will incur? Is there a defined process to be followed and an independent board that decides on the punishment? Or is it up to the individual teacher to decide how bad the plagiarism is and what the punishment should be? In a way, this last situation, which is quite prevalent in Germany, simultaneously forces teachers to be police (detecting the plagiarism), prosecuting attorney (putting together a case), judge (determining guilt or innocence), and enforcement officer (making sure that the punishment is carried out).

This chapter will look in more detail at the situation in a few countries other than Germany. The choice of country was determined more or less by the literature that was available.

6.1 United States of America

There are many different kinds of ternary educational systems to be found in the USA, from two-year community colleges through four-year state schools with a broad educational goal to more research-oriented private universities. Some national schools exist, primarily military academies, and many universities can be considered religious schools, as they are subsidized by a religious organization. Some schools are focused on educating local students, while others wish to attract many students from foreign countries who will have to pay stiff tuition fees. Both public and private schools are able to make up their own rules and regulations pertaining to codes of conduct, as there is no national or state-wide regulation of university system, perhaps with the exception of state universities.

Generally, however, the awareness about the problem of plagiarism and academic misconduct among both students and staff is much greater in the USA than, for example, in Germany. There are many different and widespread efforts underway to try and curb the amount of plagiarism occurring and to prevent other forms of academic misconduct.

6.1.1 Student Code of Conduct

Many schools and most universities publish a student code of conduct that can address a wide range of expected conduct when participating in school activities. This can be restricted to academic matters, but often also extends to issues such as concealed weapons (a major problem at US universities), profanity, smoking or drinking, and correct attire. The religious colleges and the military academies are particularly prone to dictating details (College Times 2010).

These codes of conduct spell out for the students exactly what is expected of them. They will also outline the procedures that will be followed when the code is violated and elaborate on the penalties that can be meted out. Some small to middle-sized universities, in general private ones, will have a detailed honor code that outlines the specifics of good academic practice, as McCabe & Pavela (2000) report.

Although most universities now have their code of conduct online, for example Massachusetts Institute of Technology (n.d.), Emory University (2013), or Stanford University (2013), some schools hand out a written copy on the first day of school to their first-year students. A number of schools have a pledge of honor that is part of their code of conduct. Students are required to affirm the pledge either orally or in writing, or both, and will also pledge to help others uphold the honor code. A video of the pledge being given orally at Vanderbilt University (n.d.) can be found on the Internet. Stipulations in the honor code can go to the extreme of examinations being unproctored or students being required to turn in cheaters in order to avoid being found guilty of academic misconduct themselves. At some institutions, when fellow students observe someone cheating, they are not expected to turn the cheater in. Instead, they are charged with speaking directly with the cheater, informing them of having violated the honor code, and requesting that they turn themselves in.

6.1.2 Honor Board

The official body that is charged with determining academic misconduct on the part of students and deciding on sanctions is often called an honor board. This body meets regularly to decide upon cases that have been submitted to them by faculty. There is often a requirement that all cases of misconduct must be submitted to this board, so that a fair and equal application of the rules can be guaranteed. This also means that teachers cannot just accuse a student of plagiarism: They must prepare a formal report with materials that the honor board will use to judge the case.

In general, the honor board members will consist of a majority of students with faculty participating to ensure continuity from year to year. A member of the president's or chancellor's office may also be a member of this body *ex officio*. At some schools a suspension for one or two semesters would be given for a first-time offender, although there are universities, such as the University of Virginia, at which the only possible sanction is to expel a student for any offense confirmed (University of Virginia 2013).

6.1.3 Sanctions

Each university has its own specific sanction catalog for students caught cheating. The University of California, San Diego has adapted its catalog from the University of California, Riverside and explicitly spells out the possibilities (University of California, San Diego 2011). UCSD defines four levels of violations and the accompanying sanctions.

- The **first level** is reserved for minor infractions from inexperienced students. Students may be warned, have a mark entered into their records, be required to attend a course or to write a paper reflecting on their actions, or have their grade lowered and not be permitted to retake the class to improve the grade.
- The **second level** is for more serious violations that include a limited amount of plagiarism. Sanctions for this level now include the student being placed on academic probation.

- The **third level** is for "flagrantly dishonest" work that is demonstrated to have been deliberately conducted and can now be sanctioned with suspension and a mark on the academic record that will be sent on a transcript to potential employers or other universities upon request.
- The **fourth level** is either for repeat offenders or for any offense committed by students at the graduate level, and can result in expulsion from school with a permanent mark that they were "suspended for academic dishonesty."

Instead of codifying the sanctions into just one PDF document put somewhere on the university web site, the university has set up an entire subsite on academic integrity that addresses students in a more informal tone (University of California, San Diego n.d.). They sum up academic integrity in the section "Consequences":

So the costs of cheating outweigh the benefits?

Absolutely! Educate yourself and your friends and always, always do your academic work with integrity!

The University of Virginia, mentioned above as a "single sanction" university (University of Virginia 2013), does not bother with defining levels of misconduct and appropriate penalties. They have only one sanction for *any* non-trivial act of lying, cheating, or stealing: dishonorable dismissal from school. This means that students at the university discuss the topic of honor and cheating very often with their professors and among each other – they want to be sure that they do not act in such a manner as to incur this drastic sanction.

6.1.4 Faculty Code of Conduct

Most schools in the USA will also have a code of conduct for faculty that will be laid out in the faculty handbook that is given to all staff members and updated as necessary, for example, the University of California, Santa Barbara (2008). These codes of conduct can also include aspects such as dress codes and outline permissible private contact with students outside of class. The faculty is considered to be informed about the processes that are in place since they are laid out in writing in the handbook, so teachers do spend time studying them.

For each course that a member of the faculty teaches, there is a syllabus that must be prepared and handed out to students at the beginning of each semester that details the amount of collaboration that is acceptable for this class, as this may vary widely. Teachers who fail to set up such a syllabus may also be sanctioned.

6.1.5 Ethics in Research

Researchers at universities in the USA are expected to attend mandatory ethical training, and in some places this training must be repeated every few years. This is especially important for research that involves human or animal subjects. There are strict regulations in place and researchers must obtain approval from an ethics board before they can proceed with their research in these fields. Not following procedures can lead to researchers being prohibited from applying for funding for summer research (and thus not having a salary for three months, which is still standard in the USA) or even fired for academic misconduct.

There is a federal organization called the *Office of Research Integrity* (ORI) that is charged with overseeing research integrity activities in biomedical science for the Public Health Service. They have developed excellent training materials that universities can use for their own ethics courses. Charges of academic misconduct can also be investigated by the ORI, which publishes the results of confirmed misconduct online.

6.1.6 Teach Writing

Many universities spend the first and second years of a bachelor's program preparing the students for university work. They are trained first in general critical thinking and writing, as well as given a footing in mathematics and other basic subjects for the major program chosen. Community colleges offer remedial writing courses that teach the basics of research, structuring a paper, and the actual writing, as well as critiquing the writing of others.

Modern students, more rooted in a visual or audio world than their instructors, often have trouble setting out a linear chain of arguments to support a position, and sometimes even have grave deficits in simple questions of spelling and grammar. There is no other way to learn writing than to write – and receive feedback from others on what is understandable and what is not.

Some universities also offer writing clinics. These are offices staffed with writing teachers who can work with students on an individual basis who are having trouble with a term paper or a thesis. The staff at writing clinics are able to navigate the fine line between giving a student advice on how to write and actually writing the paper for them.

6.1.7 Libraries in Charge

In many universities in the USA, it is the central university library that is charged with organizing both the education of teachers and students about plagiarism, and with assisting educators with a suspicion on the use of plagiarism detection software. Librarians are trained to do research, to find materials, and to deal with large quantities of information. Since plagiarism is not a problem that is specific to one particular field of knowledge, they are able to offer guidance that is independent of the specific area of content.

A good number of materials for teaching students about plagiarism have been developed by librarians and are offered online. To list just a few examples,

- The Paul Robeson Library at Rutgers University offers an amusing series of cartoon films called the "Anti-Plagiarism Game Show" (Rutgers Universities Libraries n.d.),
- The Valdosta State University Library (n.d.) offers a 12-minute movie called "Crime and Punishment"
- The Central Piedmont Community College Library (n.d.) has an interactive, animated film called "Dr. Cite Right"

Almost all libraries have at least a web page about plagiarism, and many will refer to an amusing and elaborately produced five-minute short film from the University Library in Bergen, Norway, called *Et Plagieringseventyr* (University of Bergen 2010), an adventure in plagiarism.

The libraries do not see themselves as the determiners of plagiarism. Instead, they are service centers that offer advice and training. They will often be charged with choosing, installing, and administrating a plagiarism-detection software tool that is available to staff – and at some schools even the students as a formative measure. Burke (2004) gives a good overview of the role that libraries could be playing in dealing with plagiarism.

6.1.8 Selling Term Papers

One suggestion that is often offered for curbing the ghostwriting problem is to outlaw the sale of term papers. In the USA, at least 13 states (Standler 2012, p. 68) have already made it illegal to sell term papers, theses, or dissertations to students. Standler found the first references to such laws enacted in 1972 and notes that even though it is a misdemeanor such an offense can incur between two and six months in jail and a fine of up to $1,000. Standler noted that in theory, each sale of a paper would be a separate offense, so selling 1,200 papers could theoretically put a businessperson behind bars for 200 years, although such crimes are not generally punished that hard.

Standler also noted that in a few states universities have actually sued ghostwriting businesses. For example, in *New York v. Saksniit*, 332 N.Y.S. 2d 343 (Sup. Ct. 1972), the State of New York sued Kathleen Saksniit, the owner of a paper mill, with the goal of keeping the company from committing fraudulent acts by selling ready-made term papers (Ambash 1973). The court decided that Saksniit must quit advertising, producing, or delivering term papers to students. The assets of the company were frozen, essentially prohibiting the company from continuing to do business in the state. However, shutting down one operation tends only to make them pop up elsewhere with a new name. A market exists and there is much money to be made, so this kind of business will continue to exist, no matter if they are outlawed or not.

There was apparently a problem that ghostwriters had for a while in that they were checking the papers they wrote for plagiarism with *Turnitin*. Since this caused the papers to be stored in the database and would raise a red flag when they were handed in, term paper ghostwriters will now advertise that they do not use Turnitin.

6.2 England/United Kingdom

England has a national system of universities and new universities (former polytechnics) that is federally funded by the Higher Education Funding Council for England (*HEFCE*). There are, however, an increasing number of higher education institutions that are not recognized as universities, as well as several private institutions that were granted degree-awarding powers in 2012.

6.2.1 Plagiarism Reference Tariff

In England the company that markets *Turnitin*, iParadigms Ltd., commissioned and supported studies into setting up a plagiarism reference tariff system (Tennant, Rowell, & Duggan 2007; Tennant & Duggan 2008; Tennant & Rowell 2010a,b) as an aid to determining how serious a plagiarism offense by a student is and what the penalties should be.

The tariff was set up because of concerns that the penalties applied for plagiarism were inconsistent from university to university. A project was funded, the Academic Misconduct Benchmarking Research project (AMBeR), to investigate the current situation at UK universities and set up a benchmark that can be useful to universities setting out or revising their plagiarism policy.

There were 104 persons from the 168 UK-funded institutions who were interviewed by online questionnaire. They looked at the factors that were used at the different schools for reacting to plagiarism incidents and derived the tariff from this data. Tennant & Rowell (2010a) give a detailed description of the project and the process by which they derived the tariff.

The tariff is based on a points system, with points being assigned for history (first, second, third or more offense), the extent of the plagiarism (amount plagiarized, critical section or not, or term paper purchased), level of the term paper (bachelor, master, or doctoral work), the weight of the assignment (small or large), and with extra points assigned for attempts to disguise the copying.

Penalties range from a formal warning that is also recorded in the student's records (for example, for a first offense, small assignment, less than 5% and less than 2 sentences plagiarized in a bachelor's program) over awarding 0 marks for the assignment, resubmitting for a capped grade, a re-sit for a capped grade, a reduction in the level of the degree awarded (for example, from an honors degree to a non-honors degree), to expelling with earned credit retained. The highest penalty amounts to expelling the student with all credits earned up until the present day withdrawn, and thus not transferrable. According to Tennant et al. (2007, pp. 13–14), of the 93 UK higher education institutions surveyed, there were over 9,000 cases of plagiarism and cheating reported in one year, resulting in 143 students being expelled from their institutions.

Although the Higher Education Academy recommended that universities adopt this system, they are unable to require the use of the tariff, since each university has its own local policies and procedures. It appears that the system, perhaps because of its complexity, is not being widely used. But it did foster the notion that criteria can be defined and applied for judging the seriousness of an offense.

6.2.2 *Plagiarism Detection Software*

The UK first purchased a nationwide license from iParadigms, Ltd. in 2002 for using their plagiarism detection software *Turnitin*. The company reports that 95% of universities in the UK are actively using the system (Barrie 2008). The government subsidized setting up the organization *plagiarismadvice.org* in Newcastle-upon-Tyne through the Joint Information Systems Committee (*JISC*), which is now a private company, up until 2008. It was charged with educating schools and teachers about plagiarism and on the use of the *Turnitin* system.

An international plagiarism conference has been organized and held in Newcastle, England, every other year since 2004 by *plagiarismadvice.org* and is documented on their web site. Unsurprisingly, the conference tends to focus heavily on the use of *Turnitin*. iParadigms has started marketing its product as a formative device, that is, permitting students access to the system so that they can review their work before handing them in. iParadigms is also offering an online grading and feedback system for teachers to use and a collaborative critiquing tool so that they are focusing more on learning and writing and less on cut and dried identification of plagiarists.

However, Thompsett & Ahluwalia (2010) have measured student attitudes towards the use of the system as a learning tool. The students did not find *Turnitin* easy to use and did not think that it was useful as a learning tool. Even though there were training materials available, the students did not appear to use them and were unsure what the results of the system meant. They also pointed out that *Turnitin* itself does not actually teach about good citation habits; it only points out text copies that could be classified as plagiarism.

Davis & Carroll (2009) conducted a similar study with comparable results, determining, however, that the software does wake up the students, but that the teaching and learning occur when the students speak with teachers about the reports and not just when reading the reports.

6.2.3 Oxford University

Nicholas Bamforth (2013) describes the system for combating plagiarism used at the University of Oxford in England. Two faculty members are freed from their teaching duties for an entire year in order to be in charge of dealing with cases of all academic misconduct (not just plagiarism) at the university. One, called the Senior Proctor, is responsible for the graduate students; the other, the Junior Proctor, is responsible for the undergraduates.

Bamforth elaborates on the formal procedure that is defined for filing a complaint, and notes that the proctors have the right to access any and all files that are available at the university in order to look into the complaints and decide on the consequences. All plagiarism charges must be referred to the Proctor's office and may not be handled by the individual examiners (Bamforth 2013, p. 73).

The Proctor's office is also charged with informing the new students about the code of conduct. It is available online at (University of Oxford, Proctors' Office n.d.) and handed out in printed form to new students every year. The code not only regulates general conduct and avoidance of academic misconduct; it also stipulates when caps and gowns are to be worn and to avoid champagne and silly-string near the examination halls. It also makes it clear on the web page in section "Essay-writing services" that members of the university may not ghostwrite for others[1]:

> As can be seen in the provisions of the Code of Discipline [...], no member of the University is allowed to contribute to essay-writing services (whether as part of private arrangements with the recipient or through commercial companies) in circumstances where the work provided could be submitted by someone else, at Oxford University or elsewhere, in an examination. The Proctors will treat any infringement of this regulation particularly seriously because such activities undermine students' academic work and fair examination processes. Any Oxford student who buys or otherwise obtains material to pass off as his or her own in a University Examination would be in breach of the Proctors' Disciplinary Regulations for University Examinations [. . .] and can expect to be penalised severely.

The Telegraph (Rowley 2011) reported that in the past academic year the proctory at Oxford found only 12 students guilty of academic misconduct, not including students fined for taking mobile phones into the examination room with them. Two of these students were expelled from the university; four were given no grade for their work. But the proctors note in the article that this is probably just the tip of the iceberg as only "the most brazen instances of cheating would be reported to the proctors." The article also notes that there were 17,000 cases of plagiarism and academic misconduct reported nationwide, up 50% from the year before.

Part of the problem might be found in the specific model of instruction used at Oxford. Instead of collecting credits, the students have exams at the end of years one and three – not at the end of each course. Instead, they meet with a tutor for one-on-one instruction every week, often reading a short essay out loud that they wrote the previous week and getting a new topic for the coming week. So when they are asked to write academically for a research paper, they do not have much

---

[1] Although one of the major players in contract cheating in the UK guarantees that their papers are all written by "top" Oxford or Cambridge graduates, according to their web site.

experience in doing so. This is quite similar to the situation in the natural sciences and engineering in Germany.

6.2.4 Oxford Brookes Model

Oxford Brookes is an English "new university" (i.e., former polytechnic) where Jude Carroll worked for many years. She is the author of numerous books and papers on combating plagiarism, in particular the good practice guide she wrote with Jon Appleton (2001) and her handbook for deterring plagiarism in higher education that served as an inspiration for this book (Carroll 2007). She was able to institute many of her ideas at Oxford Brookes – which in that respect, serves as a model school in the area of dealing with plagiarism.

Carroll makes it clear that the problem with plagiarism is not one of not having hard enough punishments, but that it is an educational and pedagogical problem. It is necessary to teach the students good academic practice so that they do not cheat, but above all it is also necessary to stop students bypassing the hard work of learning and of making ideas their own by finding, faking, buying, or pretending to be working. One good way to do this is by redesigning assessments.

In (Carroll 2007, p. 36) Carroll makes suggestions for designing courses in order to deter plagiarists, for example making sure that there is no way to pass the course using something that already exists, such as old exams, or submitting other's work as one's own, or the students choosing an assessment task that makes fraud easy. If the courses are designed so that each task builds on and confirms the previous ones, if student effort is observed and recorded, if they are encouraged to use online sources with proper references, and if they are given the chance to practice academic writing and get feedback on that writing, students will be much less prone to plagiarize, according to Carroll.

The handbook goes into much detail on using assessment to deter plagiarism. In particular, students need to be given specific instructions and they must sign a statement of originality. They are also required to submit draft work – although this places an addition burden of correction on the teacher – so that the process can be evaluated as well as the final product.

Carroll gives examples of using defined requirements or narrowing the specifications of the task, for example, using a particular book, or data provided by the lecturer, or one or more specific sources written within the past year. She also notes that asking students for reflective journals, for instance, can be used as evidence that the students did the work themselves.

She also suggests a random *viva*, an oral exam about the work done. If the student is not able to answer questions orally, the teacher will soon move from just checking up that they have learned the material into looking into a suspicion of cheating. This is something that in many countries would have to be regulated by law.

Oxford Brookes University uses a system of academic conduct officers that manage cases of academic misconduct (Oxford Brookes University 2011). Teachers register complaints with the officers, who investigate, interview the student, and then assign a sanction from a defined tariff. Carroll has evaluated and reported on the system, for example in (Carroll 2005) or (Macdonald & Carroll 2006).

6.2.5 Office of the Independent Adjudicator

If a student in England or Wales is not happy with the result of an investigation that a university has conducted, he or she has two possibilities. They can either appeal to the Office of the Independent Adjudicator (OIA), a private company that will look into the process that the university has followed and determine if it was done properly, or they can go to court to have the result investigated. OIA does not, however, assess an academic judgment or concern itself with admission issues.

The OIA was institutionalized by the Higher Education Act 2004. All universities must submit to a review by the OIA, but it has no power to sanction the universities (Office of the Independent Adjudicator n.d.). They are, however, able to publish detailed cases that name the university but not the student. Such case studies can be found on their home page, as well as material on encouraging good academic practice.

A student who was not happy with OIA's decision that his university had correctly followed procedures after determining plagiarism in his master's thesis sued OIA, although its judgments are supposed to be final. The Court of Appeal permitted this case to be looked at. Although there was much discussion about this case in England, the court finally dismissed the claim in May 2013, finding that the office acted correctly (Office of the Independent Adjudicator 2013).

6.3 Canada

Between January 2002 and March 2003 Hughes & McCabe (2006) used the International Center for Academic Integrity's Assessment Project, published in (McCabe, Treviño, & Butterfield 2002), to collect data from eleven Canadian institutes of higher education from five provinces and different types of institutions. Almost 15,000 students completed the questionnaire, including about 1,300 first-year students who were asked to reflect on their experiences in high school. There were different surveys for first-year students, undergraduates, graduate students, and faculty/teaching assistants.

Hughes and McCabe reported that there was considerable agreement on what exactly made up academic misconduct, except for six behaviors that students did not find problematic but teachers did (2006, pp. 7–8):

> [...] working on an assignment with others when the instructor asked for individual work, receiving forbidden help on an assignment, hiding library or course materials, fabricating

or falsifying lab data, using a false excuse to obtain an extension on a due date, and getting questions and answers from someone who has already taken a test.

Even though three-quarters of the respondents had suspected or were sure of having observed a cheating instance in the past year, most still did not think that cheating was a problem. They also reported that the most likely penalty for a student caught cheating would be a reprimand, although they would prefer for the student in question to get a failing grade.

Besides proctoring exams and regularly changing the content of the exams, teachers provided their students with information on cheating and discussed the importance of honesty and integrity with them. *Turnitin* was also used, mostly in the Arts departments (by around 40% of the teachers). However, only 11% of the faculty stated that these measures and policies actually worked. Hughes & McCabe (2006, p. 14) noted that only a third of the faculty reported having learned about the academic integrity rules for their school through discussions with other faculty, with their department chairs, or with the dean, while over half (58%) reported that they had only found the information by reading through written university material.

The general overview from Hughes and McCabe was that the situation in Canada was similar to the USA – students self-report to a significant level of questionable behavior but do not get caught. When they do get caught, there seem to be no negative consequences.

Cormier (2009) reports on the situation at the University of Toronto. There were 403 official cases of plagiarism in the 2006/07 academic year compared with only 92 a dozen years earlier. But since the student population had increased by half during that time, when calculating the result by percentages, only 0.6% of the students were caught cheating. If the report by Hughes and McCabe is generally valid (despite the study not being representative), then there are quite a number of students getting away with cheating.

The University of Toronto uses *Turnitin*, and Cormier reports that faculty like it, because they feel that this scares the students into working properly. Students did have issues with the system, however, not because of suspicions of plagiarism, but because the papers were stored in a server located in the USA, making it subject to legal regulations such as the Patriot Act. The company that markets *Turnitin* responded to these concerns by storing the Canadian papers on servers that are physically located in Canada.

As in the USA, libraries in Canada are very active in informing about how to avoid plagiarism. The Vaughan Memorial Library produced a tutorial called *You Quote It, You Note It!* (Acadia University 2008), McMaster University filmed an Academic Integrity video lecture (O'Connor & Thyret-Kidd 2005) that makes the reasons clear for why the words of others need to be acknowledged, and the Mount Royal University Library (n.d) offers an online lecture called *Cite it!*, to name only a few. The University of Alberta Libraries (2012) has set up a very extensive site that links to many available resources.

6.4 Sweden

Sweden is a society that is still quite heavily based on trust, and the Swedish higher education system highly values students' rights. The expectation is that people do not cheat because it hurts themselves and the community. There are still many people in Sweden who just do not cheat because they find it immoral. This is the reason why Swedish universities have been exceedingly slow at setting up systems for dealing with academic misconduct, as Jude Carroll explains in an interview with the Malmö University journal *Praktik&Teori* (Fredén 2009a). She is quoted as saying that Sweden is about five to ten years behind the UK or the USA in addressing this topic.

Swedish universities have started using software systems such as *Urkund, Genuine Text*, or *Turnitin* for checking student papers for plagiarism. The universities have come to realize, however, that both cases of false positives and of false negatives are extremely problematic. Courses are now being offered for students in the areas of writing, learning good study habits, and acquiring an academic habitus (Fredén 2009b).

In Sweden, universities will have a committee called the *disciplinnämnden* that is responsible for hearing disciplinary cases. These are not just academic misconduct cases, which only made up half of the cases in the 2008 national report (Lindén 2009), but there can also be cases of mobbing or other such problematic behavior brought before this body. The *disciplinnämnden* will have a majority of teachers on it, and at least one must be experienced in legal matters, but students are represented as well. They look at the cases presented to them, and decide on a sanction. They range from a warning given by the rector of the university up to suspension of the student from the university for a period of up to six months. During the suspension period, the sanctioned students are not allowed to participate in instruction or sit for examinations. In addition, the central student financing office is informed of the situation – and this means that their student financing is curtailed for this period. An example of such rules can be seen at (Malmö högskola n.d., in Swedish). Students who are unhappy with the rulings of this body have no recourse but to take the university to court.

The 2008 report included a small statistical analysis, with men cheating more than women (60:40). There is also a doctoral thesis (Nilsson 2008) that investigated the reports between 2001 and 2005. Nilsson interviewed some students who insisted that they were innocent, and spoke with teachers as well. He observed that teachers and students sometimes understand instructions quite differently. For example, if the teacher says "Write individually, but you are welcome to work together," some students are unsure what to do if they worked out an example together in a group. They wonder if each should now do an example on their own or if it is sufficient for each member to use the group work in their own report. It seems to be the case that students with low university entry grades and some exchange students have the most problems writing and land in trouble disproportionately often, according to the report.

The report for 2011, as discussed in the press, for example (Larsson 2012), shows an increase of 60% with respect to 2009 in the number of cases brought before the board. Throughout the country, 548 students were suspended in 2011. The government is planning on taking action to increase the awareness of students for the problem of academic misconduct.

But the basic problem at Swedish universities is perhaps that, similar to Germany, lawyers and judges manage the system, not teachers. Because processes are rather legalistic, it can take an enormous amount of time and a huge amount of proof to actually bring a charge of plagiarism and make it stick. This system is quite expensive because of the investments of time needed, and does not recognize that plagiarism is an academic issue. Plagiarism is also so common that it currently takes up much of the teacher's time.

There is no recognition of the range of levels of severity, and no agreement on the point at which an infraction becomes serious. This can also be seen in the Swedish word, *fusk*, which is used to describe everything from cheating on an exam using a crib sheet to plagiarizing or submitting ghostwritten papers. There are only two sanctions, either a warning or suspension, and then the students can continue their studies as if nothing had happened. So some teachers tend not to trust or use the system.

6.5 Denmark

In Denmark there are three committees on academic misconduct that are collected under an umbrella organization called UVVU, *Udvalgene vedrørende Videnskabelig Uredelighed* (Danish Ministry of Science, Innovation and Higher Education 2013). One committee is set up for health and medical science; one is for social science and humanities; the third is for natural, technological, and production sciences. There are six researchers appointed as members of the committees, and there is a common chair for all three committees. This is a high court judge, who is appointed by the Minister of Education.

The committees published guidelines for good academic practice in 2009 (Danish Committees on Scientific Dishonesty 2009) and publish a yearly report in Danish and in English on cases that have been presented to it. There have been a number of controversies, especially over the question of the scope of possible investigations that the UVVU can take, and about particular findings.

The report published by the Danish Ministry of Science, Innovation and Higher Education (2009) included cases of misrepresented authorship, cooking or fabricating scientific data, plagiarism, suppressing results, not obtaining proper permission for administrating, and similar problems. One of the plagiarism cases (Case 3) they reported on has an interesting twist. A hiring committee suspected that a candidate had submitted a plagiarized paper as part of the application process, and indeed found the original source after some research. They informed the UVVU about the case, and an investigation was initiated. The candidate admitted plagiarizing in responding to the allegations, but said that the purpose of the exercise was to see if anyone really looked closely at what is submitted by candidates – that is, as a test of the system. The committee was not amused, and the candidate was still found guilty of plagiarism. Sanctions can be recommended by the committee, however, it does not have any legal authority to impose them.

Danish universities invest much effort into educating their students about how to avoid plagiarism. Most universities have clear guidelines with sanctions stated that include expelling a student from the university as the highest consequence. There are procedures in place for dealing with accusations, although it can take almost a year for a case to be dealt with. Emphasis is placed on avoiding plagiarism. There is much material that has been prepared both in Danish and in English for use by educators. One example is the e-learning unit "Stop Plagiarism" (Syddansk Universitet 2010).

6.6 Finland[2]

Finland has 14 universities and 27 polytechnics (called Universities of Applied Sciences) as of 2013. According to Silpiö, all of the polytechnics publish their theses in a national open repository, *Theseus*. Many of them use either the Swedish plagiarism detection software *Urkund* or the American *Turnitin* system to check theses for crude copying. Only a few universities currently use such software, although the pressure to begin using software has been mounting, according to Silpiö, and most have announced that they will begin to use software tools in 2013.

It is difficult to find information about the situation in Finland as there are only a few English-language general interest articles and a number of Finnish-language articles on the topic. Five master's theses have been written to date about student cheating and plagiarism in Finland, as well as a few about using plagiarism detection software, according to Kari Silpiö. But next to Silpiö there is a second person active in the area in Finland, Erja Moore. She has published an article in Finnish about the culture of "silence and silencing" (Moore 2010) and an article (Moore 2013) on student plagiarism. She also blogs in Finnish about plagiarism at Plagiointitutkija (Moore n.d.).

Silpiö reports further in his master's thesis (2012, pp. 30–31) that three studies have shown that two-thirds of the students who were questioned self-reported having cheated, with 19% self-reporting having plagiarized by copy-pasting from the Internet and almost 60% self-reporting cheating on written exams. However, the studies are not statistically generalizable, as the studies used different sets of questions.

---

[2] Kari Silpiö, a senior lecturer at the Haaga-Helia University of Applied Sciences in Helsinki, kindly put together an English summary of some Finnish publications that was used as the basis for this chapter because there were no academic sources to be found in English on the plagiarism situation in Finland. His master's thesis (2012) dealt with a meta-analysis of previous studies in Finnish on student cheating and plagiarism in higher education, among other topics regarding plagiarism.

The Finnish Advisory Board on Research Integrity (2012) has published a new policy on the responsible conduct of research that details the procedures for dealing with misconduct in Finland. However, the policy is only focused on research, although it does address doctoral dissertations and lists plagiarism as one aspect of academic misconduct. This policy took effect in March 2013.

There is a Finnish Higher Education Evaluation Council that addresses quality assurance systems for higher education, but according to Silpiö they do not require universities to have any policy on plagiarism.

Silpiö notes that church historian Luukkanen published an article in 2002 describing failed doctoral dissertations on account of plagiarism in 1847 and 1853. He also found some minutes of a faculty council meeting in 2011 about a doctoral dissertation that was rescinded at the University of Eastern Finland. The dissertation consisted of five articles (a so-called cumulative dissertation), but after two articles had to be withdrawn, the doctorate was rescinded because the three remaining articles were not sufficient as the basis for a doctorate. However, there are no central statistics on either failed doctorates or rescinded doctorates on account of plagiarism in Finland.

Silpiö recommended in his thesis that the universities and polytechnica adopt a plagiarism policy that includes defining the notion, giving guidance on avoiding plagiarism, setting up a tariff for dealing with plagiarism, stipulating education goals for students and the staff, as well as gathering statistics and monitoring the implementation of the policies.

6.7 Austria

Austria, a country with a very strong affinity to using titles (Roedig 2010), has been plagued with numerous public plagiarism charges, including the 2007 discussion about the dissertation of the Federal Minister of Science and Research, Johannes Hahn. His dissertation, from the University of Zürich in Switzerland, was then examined there and the accusations were declared to be unfounded. A Vienna professor, Herbert Hrachovec, later presented a very detailed documentation of the dissertation and its sources. Hrachovec published his material online (Hrachovec 2008), including a precise analysis of the first 100 pages, but the accusations were dismissed as being "politically motivated" (Aigner 2011). In the aftermath of the public discussion, the Austrian Commission for Research Integrity (*Österreichische Agentur für wissenschaftliche Integrität*, OeAWI) was set up in 2008.

Ulrike Beisiegel, former ombud for good academic practice for the DFG in Germany, was chair of the Commission for Research Integrity from June 2009 until December 2010. The organization was set up to be an association that is not affiliated with any university or other body, with universities, research organizations, and funding bodies as members. As of 2012, OeAWI has 32 members (Männel 2012).

The structure of the organization is different from the German model. It consists of two bodies, the board and the commission. The board consists of Austrian scientists and administrators and is charged with setting up and managing the organization (Austrian Agency for Research Integrity n.d.a), which is financed by the member institutions. Sitting on the commission are six foreign scientists, one each for the areas of humanities, life sciences, medical sciences, natural and technical Sciences, law, and social sciences. As of 2013 there were five members from Germany and one from the Netherlands in the commission (Austrian Agency for Research Integrity n.d.b).

The commission, not the board, is in charge of investigating cases, which can be brought forth by any person or institution. They look into the allegations (which may only have happened less than ten years previously) and judge the nature and severity of each alleged offense. Confidentiality is expected of the members of the commission, the whistleblowers, and the persons accused, who are informed of the allegations.

A decision is first made whether or not to investigate; only then do they speak of a case that is assigned to one of the six members as a case manager. After the investigation, a written recommendation for either solving the case by means of mediation, or suggesting a course of action for the university or research organization to take is issued to the participants. The whistleblower is only informed if they are personally involved in the case (Austrian Agency for Research Integrity 2012). OeAWI has no adjudicative powers; any sanctions must be leveled by the institutions.

Since the work of the committee started, they have submitted four yearly reports in English and in German. They have also issued a statement on handling cases of plagiarism (Austrian Agency for Research Integrity 2011) and advise the Austrian universities on issues such as the use of plagiarism software for detecting plagiarism in student papers (Männel 2012). A working group on plagiarism of about 30 members from the various institutions in Austria has been set up so that they can exchange experiences and strive to develop a common policy for dealing with plagiarism. The University of Vienna, for example, checks all student papers before they are graded (Steiner 2011).

According to the OeAWI report 2012 (Austrian Agency for Research Integrity 2012, pp. 4–5), there have been 21 cases in the four years that OeAWI has been active. 16 have been completed, four are still open, and one is currently dormant. The case distribution according to field has been

- 7 cases in social sciences and humanities,
- 6 in medicine,
- 5 in life sciences,
- 2 in law, and
- 1 in natural sciences/technology.

Many of the cases involved multiple aspects of academic misconduct. The report notes that there were nine cases involving plagiarism, four with falsification of data, eight authorship conflicts, five cases in which a senior researcher took authorship for a publication by a junior researcher or a student, and three cases of interference of research. Ten of the 16 cases completed have been confirmed to be academic misconduct by the commission. The report also observes (2012, p. 2) that the cases in which the commission has been involved with have shifted from a focus on plagiarism toward other forms of academic misconduct.

It could seem that with this commission, Austria would be a model country in dealing with cases of plagiarism. Not all would agree with this statement. One researcher, Stefan Weber, has been documenting cases of plagiarism in Austria since he discovered his own dissertation to have been plagiarized by someone in Tübingen (Humberg 2005). Working as a journalist and private scientific investigator, he has documented 38 cases of plagiarism in Austria and informed the universities, according to a radio interview (Austria Presse Agentur 2007). In the interview Weber noted that only the University of Vienna reacted to his information, usually within a day. The others either did not react or called him nasty names in public. Weber sums the situation up in the interview, noting that apparently his attempts to clarify the situations and to work for quality assurance were felt to be massive harassment on the part of the universities. Weber has now moved to Germany and publishes a blog on academic misconduct (Weber n.d.).

One curious case that was widely reported in the media in Germany has to do with Dominic Stoiber, the brother of Veronica S. (case *Vs*), the woman for whom VroniPlag Wiki was named. He, too, was accused of plagiarism in his dissertation. Stoiber (2010) submitted a thesis to the University of Innsbruck about work that his father, a high-ranking German politician, did. According to (Trenkamp 2013), the university examiners looked into whether or not the plagiarism fundamentally influenced the grade that he received. The explanation, as reported, states that if he would have given proper references and thus would have received a worse grade, perhaps since there would then be less original work and more compiled work from others, only then would it be considered fraud to plagiarize. Since they would have given him the same grade, however, with or without proper referencing, this particular case was not considered to be fraud. This convoluted reasoning is exceedingly hard to follow. The university refused, however, to publish any documentation explaining their reasoning for assigning the same grade, stating privacy concerns.

More recently, VroniPlag Wiki case *Rm* concerns a thesis in law that was submitted to the University of Innsbruck in 2002. The documentation on VroniPlag Wiki has identified text parallels on 68% of the pages. More interesting, however, is that the author had submitted a thesis on the same topic to a German university in 2000, where it was rejected there on the grounds of plagiarism (VroniPlag Wiki 2013).

6.8 Poland

Poland has instituted a system whereby all theses are checked by a plagiarism detection system before they are defended. If the software returns a score of more than 20% suspected plagiarism, then the thesis cannot be defended. There are few cases that are discussed in English that are available online. But one case that was discussed extensively is one in which the university disciplinary committee at the Nicolaus Copernicus University in Toruń, Poland, initiated a process that led to a court ruling in a plagiarism case to fine the plagiarist, who held a doctorate in English studies (Rzeczpospolita 2009), making him pay the annual salary to the person he plagiarized from, and to prohibit him from taking any academic position for eight years throughout Poland. There are apparently similar cases being prosecuted in Łódź, Warsaw, and Kraków. Plagiarism is also listed as a criminal offense in Poland with a possible sentence of imprisonment for up to three years if convicted.

The Ministry of Science and Education has purchased plagiarism detection software and made it available to the universities in Poland. In 2012, about one third of the public universities and almost all of the private universities were reported as using the software, according to Sebastian Kawczyński, the owner of the company that markets the software *StrikePlagiarism* (Kawczyński 2012).

The Polish government has also set up an open access database, *POL-on*, that is charged with collecting materials about higher education in general and that will also be collecting statistics on the prevalence of plagiarism in the future.

6.9 Slovak Republic

Július Kravjar[3] (2012a; 2012b) reported at a conference in Warsaw in 2012 and at the 5th international Plagiarism Conference in Newcastle, UK, on the national system of plagiarism detection and national central repository of theses and dissertations that was instituted in the Slovak Republic. A software for detecting plagiarism and for a central repository of theses was developed in 2009/2010 by the Slovak company SVOP Ltd.

Plagiarism seemed to be spreading in Slovakia, caused by the increasing number of higher education institutions with an ever increasing number of students. With the availability of the Internet and only a minimal student grasp of copyright and intellectual property rights issues, it became clear that testing was necessary. The reason given for developing their own system, as opposed to purchasing one already developed, was because of specifics of the Slovak language. It is highly inflected and uses many diacritical marks, which makes it difficult to use software that only looks for a 1:1 correspondence and ignores diacritics.

In 1989 there were 13 public universities and around 63,000 students in the Slovak Republic according to Kravjar. At the end of 2011 there were 20 public, 12 private, and three state-run universities, and in addition, there were four institutions that belonged to private foreign universities. The government has no control over the latter. The number of students has now increased to 250,000.

Questions concerning collecting of all theses and dissertations in an electronic form and, above all, dealing with plagiarism matters have been a recurring discussion topic within the academic community for years, but without any significant progress for many years. A first higher education institution began using a plagia-

---

[3] I am indebted to Július Kravjar from the Slovak Center of Scientific and Technical Information (Slovak Center of Scientific and Technical Information n.d.) for reviewing this chapter and supplying additional statistics.

rism detection system in 2001, but it was a lone runner until in 2008 a second university and in 2009 a third one began using software to test all student papers.

The Slovak Rector's Conference began discussing the plagiarism issue in September 2006, when two documents related to the academic ethics were approved, according to Kravjar. One was concerned with students, called "Measures to Reduce the Ethical Violations of Standards for Preparation and Presentation of the Bachelor's, Master's and Dissertation Theses" and the other was the "Code of Ethics for Higher Education Institutions Employees." The proposed measures to eliminate plagiarism did not, however, take effect. In February 2008, the Conference revisited the issue of plagiarism and asked the Ministry of Education to coordinate the activities, especially those related to the acquisition of the plagiarism detection system, and it was recommended that higher education institutions amend their regulations in such a way that includes a penalty for plagiarism.

Since there is at present no legal way to rescind a degree that has been granted, the Slovak Republic is investing in discovering plagiarists beforehand. As most of the theses are submitted in the Slovak language, there is not a pressing need seen to deal with ghostwriting yet, as the market in papers available is mostly in English or German, Kravjar notes in (2012b).

It had been planned to start checking all bachelor's and master's theses produced at Slovak universities for plagiarism in the academic year 2009/10. Dissertations and habilitations were also to be checked using the centralized system with copies of the theses stored in the national repository. But the Slovak parliament did not pass the law until October 2009, so it only went into effect in 2010.

It is obligatory for all higher education institutions operating under the Slovak legal order (35 at the end of 2011) to use the system. Before a defense, a digital copy of the thesis is forwarded to the repository for checking and storage. The thesis and the relevant metadata are kept in the central repository for a period of 70 years from the date of registration. The Ministry of Education, Science, Research and Sport manages the central repository; its operation is delegated to an institution directly managed by the Ministry.

There is also a seasonal problem – most of the theses are finished in April and May, so there are an enormous amount of tests that need to be done during these months – over 25,000 papers a month – whereas in November and December almost nothing needs to be checked. Theoretically, it takes less than five seconds to execute the originality check, including conversion, detection, and generation of the output report so the daily and monthly capacity is still sufficiently high. They manage a response time of 48 hours maximum (there is a contractual agreement on this point) and have already dealt with almost 5,000 papers a day as a peak. In two years of operation, they have already created a storage need of 3.5 TB. A group of four to five persons takes care of the system's operation as a part of their full-time duties.

Unfortunately, they do not collect statistics on the prevalence of plagiarism detected because they do not examine the reports, which are handed back to the higher education institutions for the final decision on whether or not it is a plagiarism. The Slovak Rector's Conference President Libor Vozár has said, according to (Kravjar 2012b, p. 7), that "[t]he launch of the system had mainly psychological effect – the students were more responsible in writing their work and more careful in the use of resources."

The higher education institutes do not have to pay for the use of this service as the costs of setting up and running of the system are covered by the federal Ministry of Education, Kravjar notes (2012b). He continues with the observation that a focused and properly timed educational process for the students should walk hand in hand with the implementation of advanced technologies for plagiarism detection.

Kravjar also points out that the Czech Republic has started a similar system, with 34 out of 74 higher education institutes participating in a central repository with an originality check.

6.10 The Netherlands

Universities in the Netherlands appear to use the Dutch software *Ephorus* for investigating plagiarism despite comments from den Ouden & van Wijk (2011), among others, that the system generates a very significant amount of false positives. Many universities at least have plagiarism policies in place that are communicated to the student, including penalties such as being excluded from taking exams for a certain period.

Leiden University (Leiden University n.d., p. 2), for example, states in a student information brochure:

> Plagiarism is a form of fraud and is therefore an offence. For some time now, the University has been taking active steps to combat plagiarism. Computer software is often used to analyse papers and theses. If plagiarism is proven, the relevant Board of Examiners will, as a rule, impose penalties. Their severity will depend on the seriousness of the offence, and may be influenced by previous infringements. The heaviest penalty that may be imposed is exclusion from all examinations for one full year. This might mean that you would have to wait for a year for your thesis to be marked; as a consequence, you cannot graduate during that year. The penalty may also relate to just one or a few examinations, or may apply for a shorter period.

6.11 European Union Survey

The European Union is funding a survey, Impact of Policies for Plagiarism in Education Across Europe (*IPPHEAE*), from 2010–2013 on the impact of policies for deterring and detecting plagiarism on higher educational institutes across Europe. The project partners are gathering data on each country and comparing the situations. They are also planning to undertake an evaluation of the effectiveness of the current practices and will be documenting case studies and developing new tools such as workshops and materials for helping to deal with the problem. The final reports are being published as they are finalized at (IPPHEAE 2013).

References

Acadia University, Vaughan Memorial Library. (2008). *You Quote It, You Note It!* [Online tutorial]. http://library.acadiau.ca/tutorials/plagiarism/ cited 22 August 2013.

Aigner, L. (2011). Johannes Hahn hat sich Doktortitel "erschlichen". In: *Der Standard*, 23 May. http://derstandard.at/1304552583567/Gutachten-Johannes-Hahn-hat-sich-Doktortitel-erschlichen cited 6 July 2013.

Ambash, J. W. (1973). Trapping Term Paper Cheaters by Statute. In: *American Bar Association Journal*, Vol. 59, No. 2, pp. 162–166. http://www.jstor.org/stable/25726166 cited 29 July 2013.

Austria Presse Agentur. (2007). Salzburger Plagiatsjäger wirft das Handtuch. In: *science.ORF.at*. [Radio script], 29 January. http://sciencev1.orf.at/science/news/147066 cited 22 August 2013.

Austrian Agency for Research Integrity. (2011). *Statement of the Commission for Research Integrity on Handling Cases of Plagiarism*. http://www.oeawi.at/downloads/Stellungnahme_Plagiate_April2011_e.pdf cited 6 July 2013.

Austrian Agency for Research Integrity. (2012). *Annual Report 2012*. http://www.oeawi.at/downloads/Jahresbericht-2012_e.pdf cited 6 July 2013.

Austrian Agency for Research Integrity. (n.d.a). *Board*. http://www.oeawi.at/en/board.html cited 22 August 2013.

Austrian Agency for Research Integrity. (n.d.b). *Commission*. http://www.oeawi.at/en/commission.html cited 22 August 2013.

Bamforth, N. (2013). Combating Plagiarism: the Experiences at Oxford University. In: T. Dreier & A. Ohly (Eds.) *Plagiate: Wissenschaftsethik und Recht*. Tübingen: Mohr Siebeck. pp. 66–79.

Barrie, J. (2008). *Emerging educational practices for more original writing*. [Talk], presented at the Third International Plagiarism Conference, Northumbria University, Gateshead, 23–25 June.

Burke, M. (2004). Deterring Plagiarism: A New Role for Librarians. In: *Library Philosophy and Practice*, Vol. 6, No. 2. http://www.webpages.uidaho.edu/~mbolin/burke.htm cited 8 June 2013.

Carroll, J. (2005). Handling student plagiarism: moving to mainstream. In: *Brookes eJournal of Learning and Teaching*, Vol. 1, No. 2. http://bejlt.brookes.ac.uk/vol1/volume1issue2/perspective/carroll.pdf cited 12 March 2013.

Carroll, J. (2007). *A Handbook for Deterring Plagiarism in Higher Education*. 2nd ed. Oxford: Oxford Centre for Staff and Learning Development.

Carroll, J. & Appleton, J. (2001). Plagiarism: A Good Practice Guide. Oxford Brookes University & JISC. Available at http://www.jisc.ac.uk/uploaded_documents/brookes.pdf cited 12 March 2013.

Central Piedmont Community College Library. (n.d.). *Dr. Cite Right*. [Video]. http://www.cpcc.edu/library/tutorials/DrCiteRight cited 22 August 2013.

College Times. (2010). *10 Unbelievably Strict College Campuses*. In: *College Times*, 26 April. http://collegetimes.us/10-unbelievably-strict-college-campuses/ cited 3 July 2013.

Cormier, Z. (2009) Stolen Words. In: *University of Toronto Online Magazine*. Winter edition. http://www.magazine.utoronto.ca/winter-2009/u-of-t-plagiarism-academic-dishonesty-zoe-cormier/ cited 19 May 2013.

Danish Committees on Scientific Dishonesty. (2009). *Guidelines for Good Scientific Practice*. http://fivu.dk/en/publications/2009/files-2009/guidelines-for-good-scientific-practice.pdf cited 3 July 2013.

Danish Ministry of Science, Innovation and Higher Education. (2009). *The Danish Committees on Scientific Dishonesty Annual Review 2009*. http://fivu.dk/en/publications/2010/files-2010/the-danish-committees-on-scientific-dishonesty-annual-review-2009.pdf cited 3 July 2013.

Danish Ministry of Science, Innovation and Higher Education. (2013). *The Danish Committees on Scientific Dishonesty*. [Web site]. http://fivu.dk/en/research-and-innovation/councils-and-commissions/the-danish-committees-on-scientific-dishonesty/the-danish-committees-on-scientific-dishonesty cited 18 August 2013.

Davis, M. & Carroll, J. (2009). Formative feedback within plagiarism education: Is there a role for text-matching software? In: *International Journal for Educational Integrity*, Vol. 5, No. 2, pp. 58–70. Available at http://www.ojs.unisa.edu.au/index.php/IJEI/article/view/614/471 cited 3 February 2013.

Emory University Office of Student Conduct. (2013). *Undergraduate Code of Conduct*. [Web page]. http://conduct.emory.edu/policies/code/index.html cited 22 August 2013.

Finnish Advisory Board on Research Integrity. (2012). *Responsible conduct of research and procedures for handling allegations of misconduct in Finland. Guidelines of the Finnish Advisory Board on Research Integrity 2012*. [Web page]. http://www.tenk.fi/sites/tenk.fi/files/HTK_ohje_2012.pdf cited 3 July 2013.

Fredén, J. (2009a). Det svenska fusket kan motverkas bättre. In: *Praktik&Teori*, No. 2, Tema: fusk. [Special issue of the Malmö högskola magazine on cheating], pp. 7–12. Available at http://www.mah.se/upload/GemensamtVerksamhetsstod/Kommunikationsavdelningen/Praktik_Teori/PoT-nr2-2009.pdf cited 12 March 2013.

Fredén, J. (2009b). "Varför ska bara akademikerbarn kunna normerna?" In: *Praktik&Teori*, No. 2, Tema: fusk. [Special issue of the Malmö högskola magazine on cheating], pp. 13–18. Available at http://www.mah.se/upload/GemensamtVerksamhetsstod/Kommunikationsavdelningen/Praktik_Teori/PoT-nr2-2009.pdf cited 12 March 2013.

Hrachovec, H. (2008). Die Begutachtungsfrist und der Minister. In: *Quatsch*, [Blog entry], 14 August. http://phaidon.philo.at/qu/?p=348 cited 6 July 2013.

Hughes, J. M. C. & McCabe, D. L. (2006). Academic Misconduct within Higher Education in Canada. In: *Canadian Journal of Higher Education / La revue canadienne d'enseignement supérieur*, Vol. 36, No. 2, pp. 1–21.

Humberg, K. (2005). Das Plagiat: Ein Drama in Drei Akten. In: *ZEIT Wissen*, No. 5, pp. 50–54. Available online as "Abgeschrieben und erwischt: Der Plagiator" at http://www.spiegel.de/unispiegel/studium/abgeschrieben-und-erwischt-der-plagiator-a-382779.html cited 22 August 2013.

IPPHEAE (Impact of Policies For Plagiarism in Higher Education Across Europe). (2013). Project results: National reports. [Web page]. http://ippheae.eu/project-results cited 23 December 2013.

Kawczyński, S. (2012). Ogólnopolski system antyplagiatowy. Conference: *The effective anti-plagiarism policy. The models of the institutional solutions.* Warsaw, 9 May. [Presentation slides]. https://www.plagiat.pl/cms_pdf/prezentacja%20dr%20Sebastian%20Kawczynski%20-%20Skuteczna%20polityka%20antyplagiatowa.pdf cited 5 October 2012.

Kravjar, J. (2012a). Central Repository of Theses and Dissertation and Plagiarism Detection System on a National Level in Slovakia. Conference: *The effective anti-plagiarism policy. The models of the institutional solutions.* Warsaw, 9 May. [Presentation slides]. https://www.plagiat.pl/cms_pdf/prezentacja%20Julius%20Kravjar%20-%20Central%20Repository%20of%20Theses%20and%20Dissertations%20and%20Plagiarism%20Detection%20System%20on%20a%20National%20Level%20in%20Slovakia.pdf cited 12 March 2013.

Kravjar, J. (2012b). Barrier to Thriving Plagiarism. In: *Fifth International Conference on Plagiarism*, Gateshead, Newcastle upon Tyne, UK, 16–18 July. Available at http://archive.plagiarismadvice.org/documents/conference2012/finalpapers/Kravjar_fullpaper.pdf cited 22 August 2013.

Larsson, M. (2012). Skärpt ton mot plagiat. In: *Skånskan*, 19 March. Available at http://www.skanskan.se/article/20120319/LUND/120319779/-/skarpt-ton-mot-plagiat cited 12 March 2013.

Leiden University. (n.d.). *Plagiarism.* [Web page]. http://media.leidenuniv.nl/legacy/plagiarism.pdf cited 3 July 2013.

Lindén, E. M. (2009). De kartlägger studentfusket. In: *Praktik & Teorie*, No. 2, Tema: fusk. [Special issue of the Malmö högskola magazine on cheating], pp. 17–24. Available at https://www.mah.se/upload/Gemensamt%20verksamhetsst%C3%B6d/Kommunikationsavdelningen/Praktik_Teori/PoT-nr2-2009.pdf   cited 22 August 2013.

Macdonald, R. & Carroll, J. (2006). Plagiarism – a complex issue requiring a holistic institutional approach. In: *Assessment & Evaluation in Higher Education*, Vol. 31, No. 2, pp. 233–245.

Malmö högskola. (n.d.). *Disciplinärenden.* [Web page]. http://www.mah.se/medarbetare/Juridiska-fragor/Disciplinarenden/ cited 30 July 2013.

Männel, D. (2012). *Das österreichische Modell.* [Video], Talk given at the DFG Ombudsman conference, 8 November, Bonn. Available at http://www.uni-bonn.tv/podcasts/20121108_MI_Ombudsman-Maennel.mp4 cited 6 July 2013.

Massachusetts Institute of Technology. (n.d.). *Academic Integrity at MIT: A Handbook for Students.* [Web site]. http://integrity.mit.edu/ cited 22 August 2013.

McCabe, D. L. & Pavela, G. R. (2000). Some Good News about Academic Integrity. In: *Change.* Vol. 32, No. 5, pp. 32–38.

McCabe, D. L., Treviño, L. K., & Butterfield, K. D. (2002). Honor codes and other contextual influences on academic integrity: A replication and extension to modified honor code settings. In: *Research in Higher Education*, Vol. 43, No. 3, pp. 357–378.

Moore, E. (n.d.). *Plagiointitutkija. Tarinoita plagioinnista, vilpistä ja huijauksista, korkeakoulutuksen ja tieteen harmaanmustasta alueesta.* [Blog]. http://plagiointitutkija.blogspot.fi/ cited 22 August 2013.

Moore, E. (2010). Plagiointia Suomen korkeakouluissa? [Plagiarism in Finnish Higher Education?] In: *Kever-Osaaja* Vol. 1, No. 2. Available at http://www.uasjournal.fi/index.php/K-O/article/viewFile/1230/1142 cited 10 July 2013.

Moore, E. (2013). Sloppy Referencing and Plagiarism in Students' Theses. In: *IPPHEAE conference "Plagiarism across Europe and Beyond".* 12–13 June, Brno, Czech Republic. http://ippheae.pefka.mendelu.cz/files/prezentace/we1140_Moore.pdf cited 22 August 2013.

Mount Royal University Library. (n.d.). *Cite it!* [Online Lecture] https://breeze.mtroyal.ca/visualizingcitation/ cited 22 August 2013.

Nilsson, L.-E. (2008). *"But Can't You See They are Lying?" Student Moral Positions and Ethical Practices in the Wake of Technological Change.* [PhD thesis], Gothenburg University, Sweden. Available at http://hdl.handle.net/2077/17249 cited 13 July 2013.

O'Connor, M. & Thyret-Kidd, A. (2005). *McMaster University: Academic Integrity Video.* [Video]. http://www.mcmaster.ca/academicintegrity/video/video3.html cited 22 August 2013.

Office of the Independent Adjudicator. (n.d.). *About us.* [Web page]. http://www.oiahe.org.uk/about-us.aspx cited 22 June 2013.

Office of the Independent Adjudicator. (2013). *Press Notice: High Court Dismisses Claim in Plagiarism Case.* [Press release], No. 926, 23 May. Available at http://oiahe.org.uk/media/88128/pn-mustafa.pdf cited 22 June 2013

Ouden, H. den & Wijk, C. van. (2011). Plagiarism: Punish or Prevent? Some Experiences With Academic Copycatting in the Netherlands. In: *Business Communication Quarterly.* Vol. 74, No. 2, pp. 196–200 [originally published online 13 April]. Available at http://dx.doi.org/10.1177/1080569911404405 cited 3 July 2013.

Oxford Brookes University. (2011). *Academic Conduct Officers.* [Web page] http://www.brookes.ac.uk/aske/strandOne/strandOneACO.html cited 3 July 2013.

Roedig, A. (2010). »Grüß Gott, Herr Kardinal«: Österreich liebt die Titel, zumindest vordergründig In: *Gegenworte*, No. 24, pp. 43–45. http://edoc.bbaw.de/volltexte/2011/1982/pdf/13_Roedig.pdf cited 6 August 2013.

Rowley, T. (2011). Warning over cheating at Oxford University. In: *The Telegraph*, 28 March. Available at http://www.telegraph.co.uk/education/educationnews/8409585/Warning-over-cheating-at-Oxford-University.html cited 18 March 2013.

Rutgers Universities Libraries. (n.d.). *Anti-Plagiarism Game Show: The Cite is Right.* [Online tutorial]. http://library.camden.rutgers.edu/EducationalModule/Plagiarism/citeisright.html cited 22 August 2013.

Rzeczpospolita. (2009). Plagiarism Under the Microscope. In: *Newzar, Latest News from Poland*, [Blog]. Translated by Ewelina Niekrasz. https://newzar.wordpress.com/2009/06/27/plagiarism-under-the-microscope/ cited 3 July 2013.

Silpiö, K. (2012). *Opiskeluvilppii ja plagiointi korkeakoulujen opintosuorituksissa. Kirjallisuuskatsaus ja käsiteanalyysi. Ammattikavatuksen pro gradututkielma.*

[Master's thesis, Tampere University]. http://tutkielmat.uta.fi/tutkielma.php?id= 22244 cited 30 March 2013. [Silpiö gave the author a short synopsis of this master's thesis in English].

Slovak Center of Scientific and Technical Information. (n.d.). *Theses and dissertations*. [Web page]. http://www.cvtisr.sk/en/support-of-science/theses-and-dissertations.html?page_id=789 cited 30 June 2013.

Standler, R. B. (2012). *Plagiarism in Colleges in USA: legal aspects of plagiarism and academic policy*. [Web page]. http://www.rbs2.com/plag.pdf cited 1 November 2012.

Stanford University Office of Community Standards. (2013). *Honor Code*. [Web page]. http://studentaffairs.stanford.edu/judicialaffairs/policy/honor-code cited 22 August 2013.

Steiner, M. (2011). Abgeschrieben und erwischt: So funktioniert die Plagiatsprüfung. In: *UNI:VIEW Magazin*, 18 March. Available at http://medienportal.univie. ac.at/uniview/uni-intern/detailansicht/artikel/abgeschrieben-und-erwischt-so-funktioniert-die-plagiatspruefung/ cited 6 July 2013.

Stoiber, D. (2010). *Die Föderalismusreform I der Bundesrepublik Deutschland: Beschreibung und Bewertung der Reform und eine Analyse der Bewährung in der Praxis anhand des Nichtraucherschutzes*. [PhD Thesis], Innsbruck, Austria. Abstract available at http://data.onb.ac.at/rec/AC07806833 cited 22 August 2013.

Syddansk Universitet. (2010). *stop plagiarism*. [Online tutorial]. http://en. stopplagiat.nu/ cited 22 August 2013.

Tennant, P. & Duggan, F. (2008). *Academic Misconduct Benchmarking Research Project: Part II – The Recorded Incidence of Student Plagiarism and the Penalties Applied*. The Higher Education Academy & JISC. Available at http:// www.heacademy.ac.uk/assets/documents/AMBeR_PartII_Full_Report.pdf cited 10 May 2013.

Tennant, P. & Rowell, G. (2010a). *Benchmark Plagiarism Tariff*. Available at http: //archive.plagiarismadvice.org/BTariff.pdf cited 10 May 2013.

Tennant, P. & Rowell, G. (2010b). *Plagiarism Reference Tariff*. Available at http:// archive.plagiarismadvice.org/documents/AMBeR%20Tariffv2.pdf cited 10 May 2013.

Tennant, P., Rowell, G., & Duggan, F. (2007). *Academic Misconduct Benchmarking Research Project: Part I – The Range and Spread of Penalties Available for Student Plagiarism among UK Higher Education Institutions*. Available at http://archive.plagiarismadvice.org/documents/amber/FinalReport.pdf cited 10 May 2013.

Thompsett, A. & Ahluwalia, J. (2010). Students Turned Off by Turnitin? Perception of Plagiarism and Collusion by Undergraduate Bioscience Students. In: *Bioscience Education*, Vol. 16, No. 3. Available at http://journals.heacademy.ac.uk/ doi/full/10.3108/beej.16.3 cited 30 March 2013.

Trenkamp, O. [otr]. (2013). Dissertation über Arbeit des Vaters: Stoiber-Sohn darf Doktortitel behalten. In: *Spiegel Online*, 8 May. http: //www.spiegel.de/unispiegel/jobundberuf/uni-innsbruck-dominic-stoiber-darf-doktortitel-behalten-a-898831.html cited 22 August 2013.

University of Alberta Libraries. (2012). *Guide to Plagiarism*. [Web site]. http://guides.library.ualberta.ca/plagiarism cited 22 August 2013.

University of Bergen Library. (2010). *Et Plagieringseventyr*. [Video]. http://www.youtube.com/watch?v=Mwbw9KF-ACY cited 22 August 2013.

University of California, San Diego Academic Integrity Office. (n.d.). *Working to ensure academic integrity at UCSD*. [Web site]. http://students.ucsd.edu/academics/academic-integrity/index.html cited 22 August 2013.

University of California, San Diego Academic Senate Committee on Educational Policy. (2011). *Academic Integrity Violations: The UC San Diego Response*. [Web page]. http://students.ucsd.edu/_files/Academic-Integrity/Sanctioning-Guidelines.pdf cited 3 July 2013.

University of California, Santa Barbara Academic Senate. (2008). *Bylaws and Regulations: Faculty Code of Conduct and University Policy Summaries*. [Web page]. https://senate.ucsb.edu/bylaws.and.regulations/faculty.code.of.conduct/ cited 3 July 2013.

University of Oxford, Proctors' Office. (n.d.). *Essential information for students > Proctors' and Assessor's Memorandum > Section 10: Conduct* http://www.admin.ox.ac.uk/proctors/info/pam/section10/ cited 22 August 2013.

University of Virginia. (2013). *Honor Committee Constitution*. http://www.virginia.edu/honor/wp-content/uploads/2013/03/Constitution-as-of-March-3-2013.pdf cited 24 July 2013.

Valdosta State University Library. (n.d.). *Crime and Punishment*. http://cinema.valdosta.edu/asxgen/rdevane/library_videos/crime_punishment.wmv cited 22 August 2013.

Vanderbilt University. (n.d.). *Honor Code Signing Ceremony*. [Web video]. http://blip.tv/studentvu/honor-code-signing-ceremony-68584 cited 3 July 2013.

VroniPlag Wiki. (2013). *Rm/HU: Erste Einreichung der Arbeit an der Humboldt-Universität zu Berlin*. [Web page]. http://de.vroniplag.wikia.com/wiki/Rm/HU cited 22 August 2013.

Weber, S. (n.d.). *Blog für wissenschaftliche Redlichkeit*. [Blog]. http://plagiatsgutachten.de/blog.php/ cited 22 August 2013.

Chapter 7
Outlook

The purpose of this book has been to collect material about various aspects of plagiarism in German academia. On account of the discussion surrounding the extensive plagiarism in dissertations at universities throughout the country and across all fields, the general public and the academic leadership in Germany has just now begun to realize that the problem of plagiarism must be swiftly dealt with. The German Association of University Professors and Lecturers (DHV) issued a strongly worded suggestion to its members that they not wait any longer but must get active (Deutsche Hochschulverband 2013). The DFG (German Research Foundation) published an online press packet on the future of the academic system in Germany and also addressed the problem of plagiarism and whistleblowing (Deutsche Forschungsgemeinschaft 2013). But the problem has been under discussion for over a decade now.

Since there has been much speculation in the German and international press about the GuttenPlag Wiki and the VroniPlag Wiki group, a description of the swarm phenomenon of public plagiarism documentation was included. A small selection of historical cases was also presented to demonstrate that plagiarism is not a new problem and it has not just become acute since the advent of the Internet. Techniques for discovering and dealing with plagiarism were presented, along with a selection of procedures and policies from other countries.

Where do we go from here?

Germany does not even have an agreed upon definition of plagiarism, and there is often no awareness among students and faculty alike about the various types of plagiarism above and beyond simple copy & paste. Sometimes plagiarism is simply equated with using the wrong citation scheme or even with copyright violations, which are entirely different matters. A text can, for example, be in the public domain and it thus not violate copyright to use it, but it would still be plagiarism if the text were to be incorporated in a scientific work without attribution. On the other hand, there are copyright violations that would not be considered plagiarisms, as the source is given, but the amount of text taken far exceeds the small amount of permissible use.

What seems astonishing is that so many plagiarized dissertations in Germany have been published by such esteemed publishers and have at times been given top grades by good universities. What has gone wrong? There are so many checks and balances, rules and regulations that are supposed to guarantee good quality in scientific and academic work. The only problem is, they do not work 100% of the time – there is always some charming person that can work the system to their advantage. And as water seeks the path of least resistance to flow, people seek to lessen their workload by placing their trust in others.

A good-looking, competent, high-ranking politician must have written a very good dissertation. The professor does not have to read it that carefully because he trusts his student, it is surely a *summa cum laude* thesis. The publisher sees a thesis from a well-known professor by a well-known person that has been given a good grade, and publishes the volume. Perhaps at some time in the past the publishing houses employed editors who read the texts, carefully scrutinizing every line and checking that the text made sense. But these times are long past, as publishers seek to pare their costs. People buy the book, since it is from a good publisher, and are sure that the book must be of high quality. But it has slipped through the cracks, and it is not just a singularity. The current academic administration system rewards quantity, not quality.

The desire for some sort of magical software system that would lighten the burden of dealing with plagiarism on the part of teachers and publishers is understandable, but such software can only be one tool in the toolbox used as part of the process of discovering and documenting plagiarism. General use of such software to check all student papers submitted is problematic because of both false positives and false negatives that are produced. Using them on dissertations is even more problematic, as due to the sheer size of theses other than medical dissertations, only a fraction of the text can be examined. Hoping that plagiarism will somehow go away just because a school is using software is naïve – plagiarism cannot be exterminated, no matter how much effort is expended. We can only hope to contain it.

One exception is the use of software for detecting collusion (near-duplicate papers) submitted by students in the same class. That is a task that some systems can actually master, as they can easily compare each paper submitted with each other, although the complexity of this task rises sharply as the number of student papers so examined increases. Outside of this use, there is no easy solution for teachers. This book has tried to give some practical help in the area of finding sources for plagiarism, as well as to demonstrate how easily one can use simple search engines to help discover online sources.

But the focus of the efforts on the part of the universities must clearly be in plagiarism avoidance. Plagiarism will continue to grow if there are no discussions with students and among researchers about good academic practice and no consequences following a discovery of plagiarism. If the universities do not formulate and communicate to their members a clear policy on plagiarism and academic misconduct, they will soon be in a situation where they will be unable to keep to international academic standards. People are rising through the ranks who have not learned the basics of good academic practice. They will be accepting substandard work, and perhaps continuing cutting corners themselves. It is imperative that a culture of quoting be inculcated in all participants in the academic process.

Germany is a country that has been seen internationally as an excellent research country, with much attention to detail and exact procedures being followed. The reputation of German research has been seriously harmed in recent years, but this is not the fault of those calling attention to the misconduct. It is the problem of those who do not have the courage to stand up and say: This is the line between acceptable and unacceptable academic behavior. It is the problem of conflict-averse members of academia who keep silent, or who participate in silencing actions. It is the problem of those who, for whatever reason, tolerate scientific and academic misconduct for their own personal gain.

Attention must be focused on setting up environments in which good academic practice can thrive. In such an environment the focus is on training and avoidance, not on punishment and cover-ups. There must be an ethical and moral sense of academic honesty instilled in each and every member of the community that is not codified in detailed legal norms.

Should there be a statute of limitations on rescinding doctorates? The intensive debate in Germany about the many doctoral theses containing plagiarism that were defended many years ago has focused on a call for limiting the time period for a revocation of a degree. Wolfgang Löwer vs. Gerhard Dannemann (2012) debated the pro and contra of a limitation period in a German journal on university and science politics, *Forschung & Lehre*.

Despite all the legal details that can be brought forward in order to provide the underpinnings for such a statute of limitations, as has been collected, for example, by Achim Doerfer (2012), one aspect seems to be forgotten: The conferring of a degree is an administrative act, the rescinding as well. This is not something that one goes to a third body such as a court to decide and apply for, but when a university discovers that it has made a mistake because they were deceived into thinking that they had a original dissertation submitted, they have every right to withdraw the privileges that they had granted on this basis. In a way, it is much like a building permit: If the basis for granting the permit is based on a miscalculated structural analysis or even an erroneous property title, the building may have to be demolished as the building permit never really existed.

In the United States, not only doctorates but also lesser degrees are rescinded for all sorts of reasons, from not paying off the tuition fees to convictions of rape or murder to plagiarized work discovered after the fact (see Connell 2005). There is a 1986 ruling from a well-known court case that sums the situation up quite succinctly (*Waliga v. Kent State 488 N.E.2d 850* 1986, ¶ 28):

> Academic degrees are a university's certification to the world at large of the recipient's educational achievement and fulfillment of the institution's standards. To hold that a university may never withdraw a degree, effectively requires the university to continue making a false certification to the public at large of the accomplishment of persons who in fact lack the very qualifications that are certified. Such a holding would undermine public confidence in the integrity of degrees, call academic standards into question, and harm those who rely on the certification which the degree represents.

For the sake of argument, assume that someone has been conferred a doctorate with thesis plagiarism of the same order of magnitude as was found in the thesis of Karl-Theodor zu Guttenberg. Assuming that the plagiarism was not discovered until after the statute of limitations had passed, would there be no effective sanctioning possible? That person would be able to continue to teach and do research, as they legally hold a doctorate and thus would be formally qualified for such positions. This may help make it clear why a doctorate awarded in error must be rescinded: in order make it clear that this person is not qualified to teach or do independent research.

Legal aspects may be involved in plagiarism and academic misconduct cases, but it is up to the universities to decide how serious a case is and what sanctions are necessary, all the while keeping both the persons involved and the academic community at large in mind. It is, however, a mistake to try and keep cases of alleged academic misconduct secret in a misguided attempt to protect the accused. It is necessary that such cases be dealt with in a transparent and timely manner, so that people can see the consequences of the actions of others. Bear in mind that the whistleblowers are not the problem. It is the situation in which academic misconduct thrives that must be fixed.

The call for a statute of limitations in Germany is only a very recent attempt to deal with the problem of plagiarism in dissertations. As long as only scientists were speaking about plagiarism, there was much discussion in academic journals and even in daily papers about the cases, but no one suggested that the discussion stop or no sanctions be meted out only because a certain number of years had passed. It is only since politicians, perhaps craving the respect such a title brings with it, have been doing doctorates and cutting corners that this suggestion of a time limit has been seriously discussed. The author strongly suggests that this discussion be laid to rest for the reasons given above.

Within the scope of this book, it was not possible to present a one-size-fits-all answer to the question of how to deal with plagiarism in academia. But there are, indeed, some measures that can be swiftly implemented in order to get started. Universities should

- begin discussing good academic practice openly and transparently with teachers and with students;
- set up a university competence center for advising educators individually on cases of plagiarism, which is perhaps best located in the university libraries;
- offer writing assistance and courses on good academic practice (which will include plagiarism avoidance) to students; and
- take accusations of academic misconduct seriously without calling the motives of the informer into question. A serious, objective, and effective handling of such cases is in the interest of furthering science.

It is the responsibility of all participants in academic endeavors to address this question in a rational manner.

Hopefully, this book will have given active academics food for thought and some ideas for taking action locally, in their own institutions. Cultivating a sensitivity for good academic practice cannot be ordered from the top, it must grow from below.

The problem of plagiarism will not go away, not now and not in the future – it is up to each and every member of the community to get active, to start making a difference.

References

Connell, M. A. (2005). The Right of Educational Institutions to Withhold or Revoke Academic Degrees. In: *Proceedings of the 26th Annual National Conference on Law and Higher Education*, 21 February 2005, Stetson University College of Law. Available online at http://www.stetson.edu/law/conferences/highered/archive/2005/RevokeDegrees.pdf cited 18 August 2013.

Deutsche Forschungsgemeinschaft. (2013). *Jahrespressekonferenz 2013: Elektronische Pressemappe zur Jahrespressekonferenz der DFG in Berlin.* [Online press packet], 4 July. Available at http://www.dfg.de/dfg_profil/reden_stellungnahmen/2013/130704_jahrespressekonferenz/ cited 5 July 2013.

Deutscher Hochschulverband. (2013). *AFT, Fakultätentage und DHV legen Maßnahmenkatalog vor: Konsequenzen aus Plagiatsfällen.* [Press release], 21 May. Available at http://www.hochschulverband.de/cms1/pressemitteilung+M5be7f0bf149.html cited 30 May 2013.

Doerfer, A. (2012). Die Verjährung im Promotionsrecht. In: *Wissenschaftsrecht*, Vol. 45, No. 3, pp. 227–247.

Löwer, W. vs. Dannemann, G. (2012). Pro & Contra: Verjährungsfrist für Plagiatsvergehen? Pro: Wolfgang Löwer, Contra: Gerhard Dannemann. In: *Forschung & Lehre*, Vol. 19, No. 7, pp. 550–551. Available at http://www.forschung-und-lehre.de/wordpress/?p=11177 cited 18 August 2013.

Waliga v. Bd. of Trustees of Kent State Univ. (1986). [Legal ruling]. 5 February. Available online at https://law.resource.org/pub/us/case/reporter/F2/957/957.F2d.791.91-2067.html cited 18 August 2013.

Appendices

Appendix A
DFG Recommendations for Good Scientific Practice

The German Research Foundation (DFG) published recommendations for good scientific practice in 1998 after the Herrmann/Brach scandal (Deutsche Forschungsgemeinschaft 1998, pp. 49–67). Only the recommendations are included here; the full brochure also includes much commentary on the individual proposals. The newest, 17th recommendation was not available in English, but has been translated by the author from Deutsche Forschungsgemeinschaft (2013).

Proposals for Safeguarding Good Scientific Practice Recommendations of the Commission on Professional Self Regulation in Science Recommendations:

1. Rules of good scientific practice shall include principles for the following matters (in general, and specified for individual disciplines as necessary):
   - fundamentals of scientific work, such as
     - documenting results,
     - consistently questioning one's own findings,
     - practising strict honesty with regard to the contributions of partners, competitors, and predecessors,
   - cooperation and leadership responsibility in working groups (recommendation 3),
   - mentorship for young scientists and scholars (recommendation 4),
   - securing and storing primary data (recommendation 7),
   - scientific publications (recommendation 11).
2. Universities and independent research institutes shall formulate rules of good scientific practice in a discussion and decision process involving their academic members. These rules shall be made known to, and shall be binding for, all members of each institution. They shall be a constituent part of teaching curricula and of the education of young scientists and scholars.

3. Heads of universities and research institutes are responsible for an adequate organizational structure. Taking into account the size of each scientific unit, the responsibilities for direction, supervision, conflict resolution, and quality assurance must be clearly allocated, and their effective fulfilment must be verifiable.

4. The education and development of young scientists and scholars need special attention. Universities and research institutes shall develop standards for mentorship and make them binding for the heads of the individual scientific working units.

5. Universities and research institutes shall appoint independent mediators to whom their members may turn in conflict situations, including cases of suspected scientific misconduct.

6. Universities and research institutes shall always give originality and quality precedence before quantity in their criteria for performance evaluation. This applies to academic degrees, to career advancement, appointments and the allocation of resources.

7. Primary data as the basis for publications shall be securely stored for ten years in a durable form in the institution of their origin.

8. Universities and research institutes shall establish procedures for dealing with allegations of scientific misconduct. They must be approved by the responsible corporate body. Taking account of relevant legal regulations including the law on disciplinary actions, they should include the following elements:
   - a definition of categories of action which seriously deviate from good scientific practice (Recommendation 1) and are held to be scientific misconduct, for instance the fabrication and falsification of data, plagiarism, or breach of confidence as a reviewer or superior,
   - jurisdiction, rules of procedure (including rules for the burden of proof), and time limits for inquiries and investigations conducted to ascertain the facts,
   - the rights of the involved parties to be heard and to discretion, and rules for the exclusion of conflicts of interest,
   - sanctions depending on the seriousness of proven misconduct,
   - the jurisdiction for determining sanctions.

9. Research institutes independent of the universities not legally part of a larger organization may be well advised to provide for common rules, in particular with regard to the procedure for dealing with allegations of scientific misconduct (Recommendation 8).

10. Learned Societies should work out principles of good scientific practice for their area of work, make them binding for their members, and publish them.

11. Authors of scientific publications are always jointly responsible for their content. A so-called "honorary authorship" is inadmissible.

12. Scientific journals shall make it clear in their guidelines for authors that they are committed to best international practice with regard to the originality of submitted papers and the criteria for authorship. Reviewers of submitted manuscripts shall be bound to respect confidentiality and to disclose conflicts of interest.

13. Research funding agencies shall, in conformity with their individual legal status, issue clear guidelines on their requirements for information to be provided in research proposals on (i) the proposers' previous work and (ii) other work and information relevant to the proposal. The consequences of incorrect statements should be pointed out.

14. In the rules for the use of funds granted, the principal investigator shall be obliged to adhere to good scientific practice. When a university or a research institute is the sole or joint grantee, it must have rules of good scientific practice (Recommendation 1) and procedures for handling allegations of scientific misconduct (Recommendation 8). Institutions which do not conform to recommendations 1 to 8 above shall not be eligible to receive grants.

15. Funding organizations shall oblige their honorary reviewers to treat proposals submitted to them confidentially and to disclose conflicts of interest. They shall specify the criteria which they wish reviewers to apply. Quantitative indicators of scientific performance, e.g. so-called impact factors, shall not by themselves serve as the basis for funding decisions.

16. The Deutsche Forschungsgemeinschaft should appoint an independent authority in the form of an Ombudsman (or a small committee) and equip it with the necessary resources for exercising its functions. Its mandate should be to advise and assist scientists and scholars in questions of good scientific practice and its impairment through scientific dishonesty, and to give an annual public report on its work.

The 17th recommendation from 4 July 2013:

17. Academics who give specific information about a suspicion of scientific academic misconduct (so-called whistleblowers) must not experience disadvantages in their own scientific and professional careers. The ombudspersons and the institutions that investigate an accusation must be committed to instituting appropriate means for protecting them. The accusation must be lodged in good faith.

References

Deutsche Forschungsgemeinschaft. (1998). *Vorschläge zur Sicherung guter wissenschaftlicher Praxis. Empfehlungen der Kommission "Selbstkontrolle in der Wissenschaft". Denkschrift.* Weinheim: DFG. Available at   http://www.dfg.de/download/pdf/dfg_im_profil/reden_stellungnahmen/download/empfehlung_wiss_praxis_0198.pdf cited 5 July 2013.

Deutsche Forschungsgemeinschaft. (2013). *Jahrespressekonferenz 2013: Elektronische Pressemappe zur Jahrespressekonferenz der DFG in Berlin.* [Online press packet]. Available at http://www.dfg.de/dfg_profil/reden_stellungnahmen/2013/130704_jahrespressekonferenz/ cited 5 July 2013.

Appendix B
Franklyn-Stokes & Newstead: Cheating Behaviors

Teachers sometimes equate cheating and plagiarism. Arlene Franklyn-Stokes and Stephan E. Newstead put together a catalog of cheating behaviors (Franklyn-Stokes & Newstead 1995, Appendix 1) that demonstrate how broad the concept of cheating is, which is the most prevalent academic misconduct amongst students. This list makes it clear that there are many more problems that need to be addressed with students than only plagiarism.

- Allowing own coursework to be copied by another student.
- Taking unauthorized material into an examination (e.g. 'cribs').
- Fabricating references or a bibliography.
- Lying about medical or other circumstances to get special consideration by examiners (e.g. the Examination Board take a more lenient view of results; extra time to complete the examination).
- Copying another student's coursework with their knowledge.
- Lying about medical or other circumstances to get an extended deadline or exemption from a piece of work.
- Submitting coursework from an outside source (e.g. a former student offers to sell pre-prepared essays; 'essay banks').
- A student taking an examination for someone else or having someone else take an examination for them.
- In a situation where students mark each other's work, coming to an agreement with another student or students to mark each other's work more generously than it merits.
- Continuing to write in an examination after the invigilator has asked candidates to stop writing.
- Copying another student's coursework without their knowledge.
- Illicitly gaining advance information about the contents of an examination paper.
- Inventing data (i.e. entering non-existent results into the data base).

- Not contributing a fair share to group work.
- Ensuring the availability of books or journal articles in the library by deliberately mis-shelving them so that other students cannot find them, or by cutting out the relevant article or chapter.
- Paraphrasing material from another source without acknowledging the original author.
- Copying material for coursework from a book or other publication without acknowledging the source.
- Premeditated collusion between two or more students to communicate answers to each other during an examination.
- Copying from a neighbour during an examination without them realising.
- Altering data (e.g. adjusting data to obtain a significant result).
- Doing another student's coursework for them.
- Submitting a piece of coursework as an individual piece of work when it has actually been written jointly with another student.

References

Franklyn-Stokes, A. & Newstead, S. E. (1995). Undergraduate Cheating: Who does what and why? In: *Studies in Higher Education*, Vol. 20, No. 2, pp. 159–172.

Appendix C
Managing Cheating in the Classroom

The contents of the following two lists are taken from two tables found in a paper by Donald L. McCabe, Linda Klebe Treviño, and Kenneth D. Butterfield (McCabe, Treviño, & Butterfield 2001, Tables 3 and 4) and are used by permission. They list the actions most often suggested by students to stop them from cheating and give ten suggestions for faculty for encouraging academic integrity.

The Student's Perspective
This list was adapted from students' comments on the survey McCabe, Treviño, and Butterfield conducted in 1999; the items are listed in the order of importance.

- Clearly communicate expectations (e.g., regarding behavior that constitutes appropriate conduct and behavior that constitutes cheating)
- Establish and communicate cheating policies and encourage students to abide by those policies
- Consider establishing a classroom honor code – one that places appropriate responsibilities and obligations on the student, not just the faculty member, to prevent cheating
- Be supportive when dealing with students; this promotes respect, which students will reciprocate by not cheating
- Be fair – develop fair and consistent grading policies and procedures; punish transgressions in a strict but fair and timely manner
- When possible, reduce pressure by not grading students on a strict curve
- Focus on learning, not on grades
- Encourage the development of good character
- Provide deterrents to cheating (e.g., harsh penalties)
- Remove opportunities to cheat (e.g., monitor tests, be sure there is ample space between test takers)
- Assign interesting and nontrivial assignments
- Replace incompetent or apathetic teaching assistants

10 Principles of Academic Integrity for Faculty
These principles were first published in McCabe & Pavela (1997).

- Affirm the importance of academic integrity
- Foster a love of learning
- Treat students as an end in themselves
- Foster an environment of trust in the classroom
- Encourage student responsibility for academic integrity
- Clarify expectations for students
- Develop fair and relevant forms of assessment
- Reduce opportunities to engage in academic dishonesty
- Challenge academic dishonesty when it occurs
- Help define and support campus-wide academic integrity standards

References

McCabe, D. L. & Pavela, G. R. (1997). Ten principles of academic integrity. In: *The Journal of College and University Law*, Vol. 24, pp. 117–118.

McCabe, D. L., Treviño, L. K., & Butterfield, K. D. (1999). Academic Integrity in Honor Code and Non-Honor Code Environments: A Qualitative Investigation. In: *Journal of Higher Education*, Vol. 70, No. 2, pp. 211–234.

McCabe, D. L., Treviño, L. K., & Butterfield, K. D. (2001). Cheating in Academic Institutions: A Decade of Research. In: *Ethics & Behavior*, Vol. 11, No. 3, pp. 219–232.

Appendix D
Learning Unit about Plagiarism

The following learning unit needs at least 90 minutes of instruction time. It was adapted from Wilson & Ippolito (2007) and translated into German by Debora Weber-Wulff, HTW Berlin in 2009, and published on the German-language plagiarism portal (Weber-Wulff 2009). The adapted version has been translated into English for this volume by the author. It is targeted at first semester students at an engineering college.

1. **Learning goal**

   The students learn a definition of plagiarism and have a vague notion of where the line between plagiarism and non-plagiarism is drawn. They have learned the rules at their own schools and have learned some ideas for avoiding plagiarism.

2. **Definitions**

   (5 minutes) After a short introduction, the students are given blank cards. They are asked to please write down their own definition of what plagiarism is (anonymously). They are assured that it is okay to just submit an empty card if they are unsure.

   (10 minutes) The cards are collected, shuffled, and read aloud. Elements of a definition of plagiarism are noted on the board. The following important points should be found:

   - Presenting the words or ideas of someone else as one's own (including using a ghostwriter or purchasing a ready-made paper from a paper mill).
   - Paraphrasing – changing single words or rewriting a sentence – is still plagiarism if no source is given.
   - Taking pieces from different sources and gluing them together.
   - Internet sources are not free to use, so using them without reference is also plagiarism.
   - Only giving the sources in the bibliography and not at the place that the source is used is also plagiarism.
   - It is okay to quote – one just has to make clear where the quotation starts, where it ends, and where it is from.

3. **Where do we draw the line?**

    (20 minutes) Discuss the definitions found with the entire group. Is it sufficient? Can one tell when something becomes a plagiarism? Why can one not just use Wikipedia definitions since they are under a free license? Many students consider the Internet to be a large, self-service text store.

    (10 minutes) Explain the official definition of plagiarism that exists at your school, and the policies and procedures that are defined. If your school has no such policy, work to get one installed! Discuss with students why people plagiarize.

4. **Avoiding plagiarism**

    (20 minutes) A first overview of what scientific writing is all about should be given. In particular, focus on the workflow:

    - Active reading and taking excerpts – where is the information coming from? Can I write that in my own words, or should I note down a quotation?
    - Concept phase – While I write, have I included the references? Is my literature list complete? Who said what I am currently stating? Is it from me or from someone else?
    - Composition – Have I consulted multiple sources to see if there are perhaps different opinions? Am I truly recording my understanding of the topic?
    - Proofreading – Do not just do a spelling check. Have I really referenced everything I need to? Are all of the direct quotations correct? Have I included all of the references in the bibliography?

    (5 minutes) What exactly is a reference? A quotation? What is the difference between a direct and an indirect quotation? How are they marked? Why do we need to bother with this?

    (15 minutes) The students work in groups looking through books and journals from their chosen field of study, looking for ways of referencing. The different ways of giving a reference are collected on the board during the last 5 minutes.

5. **Summary**

    (5 minutes) Summary and questions

References

Weber-Wulff, D. (2009). Lerneinheit über Plagiat. Adaptiert aus (Wilson & Ippolito 2007). [Web page]. http://plagiat.htw-berlin.de/wp-content/uploads/lerneinheit-ueber-plagiat.pdf cited 22 August 2013.

Wilson, F. & Ippolito, K. (2007). Working Together to Educate Students. In: T. S. Roberts (Ed.), *Student Plagiarism in an Online World: Problems and Solutions*. Hershey, PA: IGI Global.

Appendix E
Introducing a Modified Honor Code

Honor code researchers Donald L. McCabe and Gary Pavela (2000) discuss the implementation of a modified honor code that is also feasible for larger universities. The major steps involved are outlined here; the paper has more detail on the topic.

- Listen to the students – ask them to explain the nature and extent of cheating that is currently happening on campus. Here a student leader group can be recruited to collect information on the nature and extent of academic dishonesty at the current time.
- Students and faculty who are troubled by academic dishonesty should be given a voice in setting campus policy.
- Students should be allowed to play a major role in the resolution of contested cases.
- Sanctions should be defined that are keyed to an academic integrity seminar. For example, they suggest introducing an XF grade on transcripts ("Failure due to academic dishonesty") that can be removed by attending a workshop on good scientific practice.
- Help student leaders to educate their peers.

The International Center for Academic Integrity (ICAI) at Clemson University (http://www.academicintegrity.org/icai/home.php) offers a wide range of resources for universities wishing to institute such policies.

References

McCabe, D. L. & Pavela, G. R. (2000). Some Good News about Academic Integrity. In: *Change*. Vol. 32, No. 5, pp. 32–38.

Appendix F
Internet Home Page Register

This table gives the home pages for the products, services, and organizations that are mentioned in this book. Listing a product here does not mean that the author endorses the use of the product; the list is provided as a service to the reader.

Table F.1: Internet Home Page Register

Entity	Internet Home Page
Abbyy Fine Reader	http://finereader.abbyy.com/
ACM	http://www.acm.org/dl/
Adobe Acrobat Professional	http://www.adobe.com/products/acrobatpro.htm
APA style	http://www.apastyle.org
arXiv	http://de.arxiv.org
Babelfish	http://www.babelfish.com
BookEye	http://www.imageaccess.de/index.php?page=ScannersBookScanner4
BOSS/Sherlock	http://www.dcs.warwick.ac.uk/boss/history.php
CiteSeerX	http://citeseerx.ist.psu.edu/
Compilatio	http://www.compilatio.net/en/
Copyscape	http://copyscape.com/
CrossCheck	http://www.crossref.org/crosscheck/index.html

Continued on next page

Table F.1 – continued from previous page

Entity	Internet Home Page
Degree Essays	formerly http://www.degree-essays.com/, now http://www.ukessays.com
DFG	http://www.dfg.de/en/index.jsp
DHV	http://www.hochschulverband.de/cms1/
DNB	http://www.dnb.de/EN/Home/home_node.html
Dropbox	http://www.dropbox.com
Ephorus	https://www.ephorus.com/
Eve 2	http://www.canexus.com/
FairUse	http://www.uni-bielefeld.de/soz/fairuse/
Freenode Channel #vroniplag	http://webchat.freenode.net/?channels=vroniplag
Genuine Text	http://www4.genuinetext.com/gt4/gtMain/gt.jsp?lang=en
Google Books	http://books.google.com
Google Scholar	http://scholar.google.com
Google Translate	http://translate.google.com
Higher Education Funding Council for England	http://www.hefce.ac.uk/
HRK	http://www.hrk.de/
IEEE	http://ieeexplore.ieee.org/Xplore/
International Center for Academic Integrity	http://www.academicintegrity.org/
Internet Archive	https://archive.org
IPPHEAE	http://ippheae.eu/
iThenticate	http://www.ithenticate.com/
JISC	http://www.jisc.ac.uk/

Continued on next page

Table F.1 – continued from previous page

Entity	Internet Home Page
JPlag	https://jplag.ipd.kit.edu//
Jurion	https://www.jurion.de
Juris	http://www.juris.de
MEDLINE	http://www.nlm.nih.gov/bsd/pmresources.html
MLA style	http://www.mla.org/style
MOSS	http://theory.stanford.edu/~aiken/moss/
OeAWI	http://www.oeawi.at/en/index.html
Office of the Independent Adjudicator	http://www.oiahe.org.uk/
Office of Research Integrity	http://ori.hhs.gov/
Ombudsman für die Wissenschaft	http://www.ombudsman-fuer-die-wissenschaft.de
OmniPage	http://www.nuance.com/for-business/by-product/omnipage/index.htm
plagiarismadvice.org	http://www.plagiarismadvice.org/
POL-on	http://polon.nauka.gov.pl/
PubMed	http://www.ncbi.nlm.nih.gov/pubmed
Retraction Watch	https://retractionwatch.wordpress.com/
SFX	http://www.exlibrisgroup.com/category/SFXOverview
SPlaT	http://splat.cs.arizona.edu/
Springer	http://www.springerlink.com
Subito	http://www.subito-doc.de/index.php
Tesseract	http://code.google.com/p/tesseract-ocr/
Thesus	http://www.thesus.fi/web/guest/
Turnitin	http://turnitin.com/

Continued on next page

Table F.1 – continued from previous page

Entity	Internet Home Page
Urkund	http://www.urkund.se/EN/
Viper	http://www.scanmyessay.com
Visual Library	http://visuallibrary.net
Web of Science	http://thomsonreuters.com/web-of-science/
Webcite	http://www.webcitation.org
Wiso-Net	http://www.wiso-net.de
Wissenschaftsrat	http://www.wissenschaftsrat.de/home.html
WorldCat	http://www.worldcat.org/
Zeutschel	http://www.zeutschel.de/en/

Index

Index of Persons

Printed in the United States
By Bookmasters